Elements of
Transistor
Pulse Circuits

Elements of Transistor Pulse Circuits

T. D. TOWERS,
M.B.E., M.A., B.SC., C.Eng., M.I.E.R.E.

LONDON
NEWNES-BUTTERWORTHS

THE BUTTERWORTH GROUP

ENGLAND
Butterworth & Co. (Publishers) Ltd.
London: 88 Kingsway, WC2B 6AB

AUSTRALIA
Butterworths Pty Ltd.
Sydney: 586 Pacific Highway, NSW 2067
Melbourne: 343 Little Collins Street, 3000
Brisbane: 240 Queen Street, 4000

CANADA
Butterworth & Co. (Canada) Ltd.
Toronto: 14 Curity Avenue, 374

NEW ZEALAND
Butterworths of New Zealand Ltd.
Wellington: 26–28 Waring Taylor Street, 1

SOUTH AFRICA
Butterworth & Co. (South Africa) (Pty) Ltd.
Durban: 152–154 Gale Street

First published in 1965 for WIRELESS WORLD
by Iliffe Books Ltd.
Second edition published in 1974
by Newnes-Butterworths, an imprint of
the Butterworth Group

© T. D. Towers, 1965, 1974

ISBN 0 408 00130 5

Printed in England by
The Whitefriars Press Ltd., London and Tonbridge

D
621.3815'34
TOW

Preface to Second Edition

In a curious way, the flood of digital microcircuits readily commercially available to the circuit designer in the 1970s has made it all the more essential that he understand the functions of the various pulse circuit blocks described in this book. This is irrespective of whether these circuits are made up from 'discrete' (i.e. separate single) transistors or are bought ready-made in microcircuit form (with all the transistors internally connected for him in one package).

Opportunity has been taken to expand and update the coverage with a survey of modern 'work horse' switching transistors and diodes, together with suggested substitutions for obsolete transistors that the reader may come across in published circuit designs.

Cambridge   T. D. Towers

Preface to First Edition

In the course of my work with Newmarket Transistors and in my lectures on transistors in a number of educational establishments, I became conscious of a need for an approach text book on the application of transistors in conventional pulse circuitry.

Adequate text books exist for valve applications in this field but the newcomer to electronics (and the old timer with valve experience) will have difficulty in finding a suitable transistor approach text. This book is therefore aimed primarily at engineers and others with a limited knowledge of transistors who are interested in, or are becoming involved in, pulse circuitry using them.

The tenor of the text has been kept deliberately practical with a minimum of mathematics. Sometimes, in the interests of simplification of the analysis of circuits, I have had to be less rigorous than I would have liked. However, I have had so many encouraging comments arising out of the series of articles in *Wireless World* in 1964 on which the book is based, that I am hopeful it will fill a gap in the technical literature at the present time.

I would like to record here my appreciation of the many constructive comments I have received from friends in the transistor 'trade', and also of the great encouragement and guidance I received from the editorial staff of *Wireless World*.

Cambridge,   T. D. Towers

Contents

1 **Semiconductor and Pulse Circuits**   1

Choosing the right transistor — Germanium p-n-p switching transistors — Germanium n-p-n switching transistors — Silicon n-p-n switching transistors — Silicon p-n-p transistors — High voltage switching transistors — Silicon and germanium transistor differences — Transistor cases — Transistor specifications and substitutions — Switching diodes — Germanium diodes — Silicon diodes — Diode substitution — The pulse circuit — Phase inverter — Emitter follower — Phase splitter — Paraphase amplifier — Differential amplifier — Operational amplifier — Sign changer — Scale changer — Phase shifter — Integrator — Differentiator — Adder — Linear sweep generators — Bootstrap circuit — Miller integrator — High input-impedance amplifier — Darlington pair — Bootstrap amplifier — Regenerative switching circuits — Blocking oscillators — Multivibrators — Astable multivibrators — Bistable multivibrator — Monostable multivibrator — Schmitt trigger.

2 **Linear Pulse Amplifiers**   30

General design problems — Interstage coupling — Transistors in linear pulse amplifiers — Pulse characteristics — Frequency characteristics — Equivalent circuits — Amplifier gain and bandwidth without compensation — Typical uncompensated linear pulse amplifier stage — Collector capacitance effect — High-frequency compensation — Negative feedback H.-f. compensation — Shunt-inductance H.-f. compensation — Series-inductance H.-f. compensation — Other H.-f. compensation techniques — Low-frequency compensation — Transistors for linear pulse amplifiers — Multistage linear pulse amplifier — High-level linear pulse amplifiers.

3 **Astable Multivibrators**   48

The multivibrator family — The astable multivibrator — Operating principle of astable multi — Oscillation frequency of astable multi — Square wave generator — Practical design of astable multi — High frequency limitations — Low frequency limitations — Synchronised astable multivibrator frequency divider.

Contents

4   Monostable Multivibrators   59

Basic collector-coupled monostable multivibrator — Typical medium-speed collector-coupled multivibrator — Typical high-speed monostable multivibrator — Typical low-speed collector-coupled multivibrator — Emitter-coupled multivibrators — Complementary-symmetry multivibrators.

5   'Eccles-Jordan' Bistable Multivibrators   71

Basic circuit of Eccles-Jordan — Basic D.C. design — Switching design — Steering circuits — Speed limitations of Eccles-Jordan — Practical examples of Eccles-Jordan bistables.

6   Waveform Shaping   84

Linear waveform shaping — High-pass RC filter (differentiator) — Low-pass RC filter (integrator) — RL linear passive networks — Non-linear passive waveform shaping — Diodes as non-linear passive waveform elements — Diode clippers (or limiters) — Unbiased clamper — Non-linear active waveform shaping — Practical use of wave shaping techniques.

7   'Pumps' and 'Schmitts'   100

Diode pumps — Transistor pump — Miller integrator pump — Applications — Schmitt trigger — Refinements — Practical Schmitt triggers — Applications.

8   Blocking Oscillators   113

Operating principle — Timing mechanisms — Switching cycle detail — Design requirements — Illustrative design for monostable saturated blocking oscillator — Saturated blocking oscillators with collector–emitter and emitter–base feedback — Non-saturated blocking oscillators — Recovery time — Reverse spike — Astable operation — Output take-off — Applications.

9   'Gates'   134

Transmission v. logical gates — Transmission gates using diodes — Transmission gates using transistors — Logical gates using triodes — Symbols for logical gates — Practical aspects of logical gates.

10   Counter/Timers (Frequency Meters)   148

Typical electronic counter/timer — Counter — Readout display — Input pulse shaper — Signal gate — Master control oscillator — Timebase dividers — Control unit — Buffer amplifier and inverter — Start reset delay.

11 Timebases (Sweep-Generators)	164

Driven non-regenerative timebases — 'Transistor switch' timebase — Miller integrator timebase — Bootstrap sweep generator — Basic trigger sweep generators — Blocking oscillator sweep generators — Multivibrator regenerative sweep generators — Sweep generator linearity considerations — Symmetrical timebases — Free-running timebases — Current timebases.

Appendix A Problems	179
Appendix B Bibliography	186
Appendix C Transistor data	187
Appendix D Diode data	192
Index	195

CHAPTER 1

Semiconductor Pulse Circuits

In the last three decades, many new electronic fields have opened up where the elementary circuits or 'building blocks' have tended to become more and more sophisticated. In the fields of computers, control systems, data processing, instrumentation, nucleonics, radar and telemetry, etc., the small-signal linear amplifier is still a fundamental, but the engineer is now expected also to have an armoury of other circuits ready to hand—particularly non-linear, large-signal pulse circuits.

This book is aimed at providing the busy engineer with a practical review of the more commonly used non-linear building blocks in their transistor versions. The treatment is practical and descriptive with as little detailed analysis and mathematics as possible.

In the twenty-five years since its invention, the transistor has virtually replaced the thermionic valve in electronic switching ('pulse') circuits. In the earlier days, most transistors were germanium p-n-p slow-speed types. As was to be expected, the change-over from the valve was developed around this basic type.

By the 1970s, however, the transistor type most commonly available had become the fast silicon n-p-n 'planar' type. Apart from replacing the negative supply rail of the p-n-p with a positive rail for the n-p-n, this development luckily did not basically change the circuitry using the transistor as a switch.

Concurrent with the changeover from germanium to silicon and from p-n-p to n-p-n, the semiconductor industry devised methods of interconnecting a number of transistors together with their associated diodes and resistors in a single silicon chip to arrive at the monolithic 'integrated circuit'. These digital i.c.'s provide the system designer

2 Elements of Transistor Pulse Circuits

with ready-made self-contained pre-tested complete subcircuits. This often saves him the need to connect up discrete transistors, diodes and resistors to achieve the same switching function.

While the digital i.c. provides an economical solution to many switching problems, the use of the discrete transistor and diode still continues widely. But emphasis is now on their use to interconnect the prefabricated i.c. function blocks and to meet switching requirements in areas (such as high voltage, high current and ultra-high speed) where commercially available digital i.c.'s have limitations.

Choosing the right transistor

Competitive semiconductor manufacturers over the years have produced a confusing multiplicity of types of switching transistors and diodes. Nowadays, fortunately, only a selected few of these have received the accolade of wide acceptance, and have become practically industry-standard workhorses, available from most manufacturers. The reader may welcome some guidance on the choosing of such devices for his designs.

Apart from choosing n-p-n (positive rail) or p-n-p (negative rail), there are three main electrical specifications of a transistor to be considered for your application—the current to be switched, the voltage to be handled and the speed of switching.

For handling currents, transistors can be classified into:

(a) Low current (up to 10 mA)
(b) Medium current (up to 150 mA)
(c) High current (up to 1 A)
(d) Very high current (up to 10 A).

Units are obtainable for switching currents up to over 100 A, but these are rather unusual!

With regard to voltage, devices are available for:

(a) Low voltage (up to 6 V)
(b) Medium voltage (up to 18 V)
(c) High voltage (up to 100 V)
(d) Very high voltage (up to 1000 V).

Voltages above 1000 V are most exceptional.

For switching speeds, since it usually takes longer to switch a transistor off than to switch it on, it is common to judge the switching speed by its 'switch-off time', T_{off}. On this basis, a useable classification is:

(a) Slow (T_{off} more than 1 μs)
(b) Medium speed (T_{off} = 0·1–1·0 μs)
(c) High speed (T_{off} = 25–100 ns)
(d) Very high speed (T_{off} = 10–25 ns)
(e) Ultra-high speed (T_{off} less than 10 ns).

It should be pointed out here that there is no reason (except possibly cost) why higher current devices should not be used for low current applications. Equally, higher voltage devices can be used for lower voltage applications. In addition a higher speed device can be substituted for a lower speed. This simplifies the setting up of a useful set of work-horse standard devices.

Germanium p-n-p switching transistors

Until the 1960s, switching circuits tended to use 'old' p-n-p germanium devices, of which some of the better known ones are tabulated below. Many of these types are now either obsolete or obsolescent. Data specifications will be found in Appendix C, together with suggested modern substitutes where available.

p-n-p germanium transistors

	Low current (10 mA)	Medium current (150 mA)	High current (1 A)	Very high current (10 A)
Slow speed ($T_{off} > 1$ μs)	OC71 AC125	OC72 2N406	AC128 AD162	ASZ16 2N1533
Medium speed (T_{off} = 0·1–1·0 μs)	OC42	ASY27 2N1305	—	—
High speed (T_{off} = 25–100 ns)	ASZ21 2N967	—	—	—

Germanium n-p-n switching transistors

N-p-n polarity switching transistors were not so common as p-n-p in germanium, but the few types listed below became well known before the silicon era arrived.

4 *Elements of Transistor Pulse Circuits*

	n-p-n germanium			
	Low current (10 mA)	Medium current (150 mA)	High current (1 A)	Very high current (10 A)
Slow speed ($T_{off} > 1\,\mu s$)	2N647	AC127	AC176	AD161
Medium speed ($T_{off} = 0\cdot 1\text{--}1\cdot 0\,\mu s$)	—	ASY29 2N1306	—	—

For data and substitutions, consult Appendix C.

Silicon n-p-n switching transistors

Nowadays, other things being equal, the tendency is to use n-p-n silicon devices for switching applications. This is because in the silicon device technology it is easier to make n-p-n transistors than p-n-p. As it happened, the opposite was the case earlier on with germanium.

Well-known 'standard' n-p-n silicon switching transistors are as follows.

	n-p-n silicon			
	Low current (10 mA)	Medium current (150 mA)	High current (1 A)	Very high current (10 A)
Slow speed ($T_{off} > 1\,\mu s$)	BC107 BC182L 2N2924	BC125	BSW66 BD135 2N3054	BDY20 BD213 BD595 2N3055
Medium speed ($T_{off} = 0\cdot 1\text{--}1\cdot 0\,\mu s$)	2N703	BSW64 2N2222	BFY50 BFR39 2N2297	BDY57 2N5886
High speed ($T_{off} = 25\text{--}100$ ns)	BSY95A 2N706	BSY39 2N834	BSX59 2N3725	BDY90
Very high speed ($T_{off} = 10\text{--}25$ ns)	—	BSX20 2N2369	BSX12 2N3426	—
Ultra high speed ($T_{off} < 10$ ns)	2N709	—	—	—

For detailed specifications, consult Appendix C.

Silicon p-n-p transistors

Sometimes, as for example in replacing obsolete germanium p-n-p devices, one can by using p-n-p silicon to avoid complete circuit redesign implied by changing polarity to n-p-n silicon. P-n-p silicon transistors are readily available for switching applications as a result of continuing technological improvements in manufacture. They can, however, be at least 25% dearer than equivalent n-p-n silicon transistors. On the other hand, silicon types, whether n-p-n or p-n-p, are very much cheaper than germanium nowadays.

Commonly used p-n-p silicon switching transistors available from more than one manufacture are listed below.

	p-n-p silicon			
	Low current (10 mA)	*Medium current* (150 mA)	*High current* (1 A)	*Very high current* (10 A)
Slow speed ($T_{off} > 1\ \mu s$)	BC177 BC154	BC257	BDX14 BD136	BD214 BD596
Medium speed ($T_{off} = 0\cdot1\text{--}1\cdot0\ \mu s$)	BSW19 BC213L	BC160 BC126	BFS95 BFR79 2N4036	BDX18 2N3792
High speed ($T_{off} = 25\text{--}100$ ns)	2N2894	BSX36 2N2907	2N3763	—
Very high speed ($T_{off} = 10\text{--}25$ ns)	BSW25	2N4207	—	—

High voltage switching transistors

All the transistors described so far are for general purpose switching, which implies relatively low voltage handling. Special devices for high voltage (i.e. 100 V and usually relatively slow speed) applications such as indicator tube driving have come into common use. A selection of some of the common types of these high voltage devices are listed immediately below. Nowadays these are always silicon.

6 *Elements of Transistor Pulse Circuits*

	High voltage transistors	
	n-p-n types	p-n-p types
Low power (up to 100 mA)	BSX21, 2N1990	BSV68, 2N3497
Medium power (up to 1 A)	BSW66, 2N4390	BLX41
High power (up to 10 A)	BDX11, 2N3442	BDX18

Silicon and germanium transistor differences

When substituting silicon transistors for obsolete germanium types, there are some points to be borne in mind.

Firstly, the input 'on' drive voltage for silicon is of the order of 1 V (0·8–1·2 V) against something like 0·3 V (0·2–0·5 V) for germanium. For most general purpose circuits this is not critical, and with direct replacement of a germanium transistor by a similar silicon transistor of the same polarity (n-p-n or p-n-p), the circuit will usually be found to work satisfactorily.

Secondly, most germanium transistors were of the alloy type, for which the maximum permissible reverse voltage on the input terminal (emitter-base reverse voltage rating) was much the same as for the collector output terminal, i.e. quite high. Silicon transistors of modern types have an emitter-base rating of typically only 5 V (3–9 V) against the 15–50 V of the old germanium alloy. Because of this, before substituting silicon for germanium, check the circuit to see that the maximum reverse voltage it can impose on the transistor input does not exceed the lower rating of the silicon device. This is particularly important in one standard circuit, the astable multivibrator, discussed in a later chapter, where each of the two transistors involved can have the full rail voltage impressed in reverse across the emitter-base junction when the transistor is switched off.

Thirdly, the junction temperature of a germanium transistor is limited to less than about 100°C. This means that considerable care must be exercised to see that the device is not overheated when soldering into circuit. The use of a 'heat shunt' (e.g. flat plier jaws closed on the wire lead between the soldering iron and the transistor case) is the recommended procedure. With metal-can silicon transistors, junction temperatures up to 200°C are practicable, and it is difficult to damage a transistor in the normal soldering operation, so that heat shunts need not be used. Plastic silicon transistors discussed below under Transistor cases, have junction temperatures limited to 150°C or so. Although they are more resistant to soldering damage than germanium, you should take more care not to overheat them in soldering than with metal-can devices.

Lastly, transistor fabrication methods have been such that, while it was difficult to make a germanium transistor with a frequency cut-off over 100 MHz, it is difficult to make a silicon transistor with a frequency cut-off less than 100 MHz. As a result, modern silicon transistors tend to be more susceptible than germanium to potentially damaging short rapid voltage spikes. Germanium tended to be too slow to respond in time and just 'soaked up' the spikes. This makes the problem of circuit designing a little more complex with high speed silicon transistors.

Transistor cases

Semiconductor devices normally come in some form of protective packâge, usually with wire or metal ribbon terminal leads for access to the semiconductor element. Both in discrete devices and in i.c.'s, there used to be many different shapes of packages and arrangement of leads. Nowadays a few package outlines have become semi-standardised. In seeking a practical knowledge of transistor pulse circuits, it is necessary to have, at least, a knowledge of the more common packages. This is particularly important in order to find a substitute for a semiconductor in a design.

The semiconductor device package can be 'hermetic' or 'plastic'. Hermetic devices have the semiconductor element completely sealed in glass or metal, with leadouts issuing through glass-to-metal seals. Hermeticity of this type is still mandatory for high reliability applications.

'Plastic' devices are encapsulated in a solid block of some form of plastic. They are a low-cost alternative to the hermetic device, and are mainly used in applications, as for example in radio and TV receivers, where long life is not the major system problem. With modern silicon semiconductor device technology improvements, plastic devices can be produced with an expected life adequate under suitable conditions for a great number of applications.

Most switching transistors are manufactured in one of a few industry-standard case styles. Below about 1 A current handling, the device usually has flexible leads; and above 1 A stout pins.

8   *Elements of Transistor Pulse Circuits*

Several countries issue standards for the more common transistor cases, but in all countries most engineers tend to use 'TO' numbers registered under the American Electronic Industries Association (EIA) in their 'JEDEC' system. Illustrations of the most common transistor outlines will be found in Appendix C. For metal-can devices, these are TO1, TO3, TO5/TO39, TO18, TO66 and for plastic TO92, TO98, TO105, TO106, TO126, TOP3 and TOP66. This limited range of outlines probably encompasses over 95% of transistors in common use for switching circuits.

Low current switches nowadays almost invariably are packaged in TO18, TO92, TO98, TO106. Medium current switches use TO5/TO39 (same except for lead length) and TO105. High current devices feature TO66, TOP66 and TO126. For very high current common cases are the metal TO3 and its plastic equivalent TOP3.

Germanium low and medium current devices still use the 'old' TO1 outline, not normally to be found with silicon devices.

Transistor specifications and substitutions

Tabular specifications for all the transistors referred to by type number in this book will be found in Appendix C, with suggested replacement types in the case of obsolete devices.

In linear amplifiers, if a transistor is replaced with another type that is 'faster' (i.e. with a higher frequency cut off) or has a different output ('collector') capacitance, r.f. instability or motorboating may be produced. In switching circuits, fortunately, this difficulty seldom arises, and one can always safely use a faster transistor. Usually, of course, a slower transistor does not switch fast enough for the circuit to work correctly. (It need scarcely be pointed out again that it is always possible to substitute a higher current or higher voltage device in a replacement without adverse effect; the only drawback is that the replacement will be more costly.)

Switching diodes

As with transistors, so with diodes has there been a rationalisation of types and packages over the years. Germanium and silicon diodes are both widely used in conjunction with transistors in pulse circuits.

In diodes, germanium has not obsoleted to the same extent as in transistors. This is partly because large-scale production mechanisation has enabled the germanium diode to remain cheaper than silicon and partly because germanium diodes have an inherently lower forward voltage drop than silicon which gives them an advantage in some circuits.

Diodes are usually mounted in small double-ended wire-lead packages from 5 to 7·5 mm in length. The commonest standard packages are probably the American JEDEC DO7, DO35 and DO41 registered outlines.

Germanium diodes are always hermetically sealed in glass or metal packages, while silicon may be found in plastic as well as metal or glass.

Diodes are classified primarily by their switching speed, for which the most critical characteristic is the "reverse recovery time" T_{rr}. This is the time the device takes to switch from forward conduction to reverse cut off. For the small switching diodes generally used in pulse circuits, the reverse recovery time can range from several microseconds down to a few nanoseconds.

Other important specifications to be considered in design work are the forward voltage drop, the reverse voltage rating, the reverse leakage current and the maximum rated forward current. To help you in this area, Appendix D sets out data specifications and package outlines for a number of widely used diodes.

Germanium diodes

Germanium diode types commonly used can be 'point contact' (the oldest), 'gold bonded' and 'junction'.

Point contact types (in which a tungsten wire point is fused into the germanium chip) are relatively slow (with reverse recovery times of the order of 1 μs upwards), of limited current handling (about 100 mA maximum), with reverse voltage ratings typically 25–100 V, and with relatively high forward voltage drop (2–3 V at 30 mA). They are widely used for non-critical switching applications, mainly because of ready availability and low cost. Point contact diodes in wide use are the OA91, OA95, AAY11 and AAY21.

Gold-bonded diodes (in which a gold wire is bonded to a germanium chip) are faster than point contact, having typically less than 100 ns reverse recovery time. Also they tend to have higher current handling ability (up to about 300 mA) and lower forward voltage drop (typically less than 1 V at 150 mA). Common gold-bonded diodes are the OA47, AAY30, AAZ13, AAZ15, AAZ17 and AAZ18. Gold-bonded diodes are primarily designed for switching circuits, and the appellation 'switching diode' is often loosely used to refer to gold-bonded types.

10 *Elements of Transistor Pulse Circuits*

For higher current handling, use is made of alloy junction germanium diodes, fabricated by alloying an indium pellet into the germanium to form a junction. Junction diodes can handle much higher currents than gold bonded or point contact, but unfortunately have become obsolete. The OA10 was the best known of these used for switching purposes. Nowadays with the OA10 unobtainable, users sometimes make do for a replacement with the collector-base junction of an audio germanium transistor such as the AC128, leaving the emitter lead unconnected or cropped off.

Silicon diodes

Some use is still made of junction silicon diodes. One of the best known of this type is the 50 V OA200 (and its high voltage 150 V OA202). These are slow speed diodes with several microseconds reverse recovery times. For modern pulse circuits they have been generally superseded by faster 'diffused' switching diodes of the types detailed below.

Silicon diodes can be grouped by switching speeds into 'slow' (T_{rr} more than 1 μs), 'medium speed' (T_{rr} = 20–1000 ns), 'high speed' (T_{rr} = 2–4ns) and 'very high speed' (T_{rr} less than 2ns).

With the OA200 becoming obsolete, slow-speed high-current circuit applications are usually covered by the ubiquitous 1 A, 50 V 1N4001. This comes in a complete range from 1N4001 to 1N4007 with voltages from 50 to 1000 V.

For medium speed applications, typical common silicon diodes are the 300 mA, 150 V type BAX16 and the 800 mA, 120 V BAX12.

For high speed applications, among the better known silicon diodes are the 150 mA, 50 V BAX13; the 225 mA, 100 V 1N914 and 1N4148; and the 600 mA, 60 V BAV10. All these switch off in less than 4 ns.

Examples of very high speed (T_{rr} less than 2 ns) diodes in silicon are the 75 mA, 25 V 1N4009 and the 75 mA, 50 V 1N4151. Finally a switch-off time of less than 1 ns is found in such devices as the 50 mA, 20 V 1N4376.

Diode substitution

Usually, in replacing a diode in a circuit design, it is feasible to use a faster, higher current or higher voltage device with no ill effect. However, silicon characteristics differ so much from germanium you should exercise great care in substituting one for the other.

THE PULSE CIRCUIT

Over the years a vast number of circuits have been worked out using the transistor as a switch, an amplifier or an oscillator.

Many of these fulfil just one single special function and have little general interest to the pulse circuit engineer. Some, however, come up again and again in slight variations of a basic circuit. For example, if one wishes to switch a device positively on or off with no possibility of lingering in any state between, the choice will be, almost inevitably, some form of the 'Schmitt trigger' arrangement discussed in Chapter 8. Alternatively, in order to design around modern digital i.c.'s, it is necessary to become an expert in the 'Eccles-Jordan' bistable multivibrator covered in Chapter 6. Indeed, in order to work at all with switching circuits, it is essential to know the basic principles and uses of certain common-denominator circuits.

The pulse circuit family is now so numerous that it is well to begin with a review survey, to have a look, so to speak, at the family group photograph to identify the main features of the various members, before going on to a detailed consideration of each individually. This brief review of the principal basic pulse circuits is given below.

Phase inverter

Transistor circuitry is by now so well established that most people will have no difficulty in recognising the common-emitter transistor amplifier appearing in Fig. 1.1. In the small-signal circuits of communications practice, its most important feature is the signal-voltage amplification from input to output. The phase change across the stage is of secondary importance. This type of amplifier also appears widely in pulse circuits, but in this field the fact

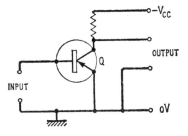

Fig. 1.1. Phase inverter

that the output is 180° out of phase with the input is much more important. So much is this so that the circuit is often known as a 'phase inverter' or simply an 'inverter'. It is frequently used merely to invert the polarity of a pulse without amplification.

Emitter follower

The other common basic amplifier in pulse circuitry is the common-collector one shown in Fig. 1.2, which is usually referred to as an 'emitter follower'. It is, of course, the transistor equivalent of the valve cathode follower. Its principal characteristics are that the output is in phase with the input, the output voltage at the emitter always lies within a fraction of a volt of the voltage at the base (hence 'emitter follower'), the input impedance is high, and the output impedance is low.

Many electronic instruments take advantage of the good stability and linearity of the emitter follower. It is usually employed where there is a requirement for a high input impedance, a low output impedance or both. The input stage of a transistor oscilloscope or electronic voltmeter is usually an emitter follower. The effect of shunt capacitance in long signal leads or screened cables can be minimised by feeding from the low output impedance of such a circuit. Where one circuit feeds into another and the reaction of the second circuit on the first is to be kept low, use can be made of an

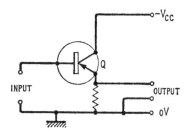

Fig. 1.2. Emitter follower

emitter follower as a buffer between the two stages. The low output resistance of the emitter follower presents a barrier to feedback from the higher input resistance of the second stage. A number of other basic circuits are described below which are similar to the emitter follower in that they have a resistor in the emitter circuit.

Phase splitter

The phase splitter (often called 'balanced inverter') shown in Fig. 1.3 provides two output signals of opposite polarity from a single input—hence the term 'phase splitter'. Output 1 (from the emitter) and the input are of one polarity, and output 2 (from the collector) is of opposite polarity. If the collector and emitter resistors are of equal value, the two voltage output signals are of equal amplitude, since by transistor action the collector and emitter currents are virtually equal. The high voltage negative feedback resulting from the emitter resistor reduces the voltage gain to just below unity at both output terminals. One defect of the circuit is that while the

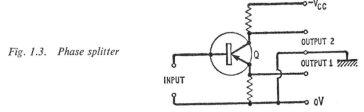

Fig. 1.3. Phase splitter

output impedance at the collector is high (approximately equal to the collector load resistance), at the emitter it is low (approximately equal to the input source resistance divided by the transistor current gain). This impedance unbalance can be got round by including in series with the emitter output a resistance equal to the collector load resistance. The two outputs are then equal in voltage, opposite in phase and from equal source impedances. They are thus completely balanced, but the circuit has no voltage gain. A typical use of this sort of circuit is to convert a single-ended sweep voltage into a symmetrical deflection for an oscilloscope.

Paraphase amplifier

Fig. 1.4 illustrates a circuit, the paraphase amplifier, which serves the same function as a phase splitter, but also provides equal signals and, without padding, equal impedances at the two outputs. This is one of the family branch known as 'long-tail pairs' or 'long-tailed pairs', which all have the same feature of an emitter resistor common to two independent transistor amplifiers through which the amplifiers react on one another. The appearance of the circuit in Fig. 1.4 with the common emitter resistor, R_E, projecting downwards makes the term 'long-tail pair' self evident. The emitters of Q1 and Q2 are close to earth potential through their forward-biased emitter-base diodes. If R_E is large, then a relatively constant direct current flows through it. If the signal voltage applied to Q1 increases, the

14 *Elements of Transistor Pulse Circuits*

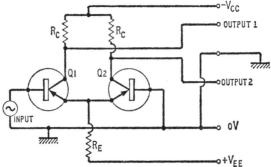

Fig. 1.4. *Paraphase amplifier*

input base current also increases. This in turn causes the emitter current of Q1 to increase. As the sum of the emitter currents of Q1 and Q2 is fixed (being equal to the total current through R_E), the current through Q2 must decrease by an amount equal to the increase in the Q1 current. These changes of current are reflected in the collector output voltages. Output 1 falls as output 2 rises. Thus for a single-ended input into the paraphase amplifier, there are available at the outputs balanced push-pull voltages from equal impedance sources (the collector resistances R_C). The circuit also provides some voltage gain from input to output.

Differential amplifier

The paraphase amplifier described above has only one input signal. It is also possible to use this basic long-tail pair to accept two signal inputs and give an output proportional to the difference between the input signals. This is the differential or difference amplifier. The basic circuit is set out in Fig. 1.5. Here the two input signals are

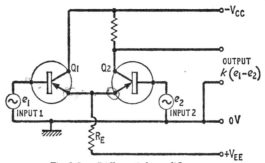

Fig 1.5. *Differential amplifier*

applied to the transistor bases. The voltage at the emitter of Q1 (and thus of Q2) follows the voltage at the base of Q1, so that the base-to-emitter voltage of Q2 is equal to $e_2 - e_1$. The corresponding signal voltage at the collector of Q2 is then by transistor action in Q2 proportional to this input voltage difference. When e_1 and e_2 are equal in amplitude but opposite in polarity, i.e. when $e_2 = -e_1 = e$, then the output of the difference amplifier, being proportional to $e_2 - e_1 = 2e$, is a single sided signal proportional to the balanced input signal. This circuit can thus be used to convert the push-pull output of a phase-splitter or a paraphase amplifier to a single-sided output with respect to earth.

In the above description, signals have been shown as a.c., but they can equally well be d.c., because all have been illustrated with d.c. coupling.

Operational amplifier

Another class of amplifiers widely used in non-linear circuits is the 'operational amplifier' shown in block form in Fig. 1.6. The internal amplifier gain A is real and negative, i.e. it has 180° phase shift and greater than unity amplification from input to output.

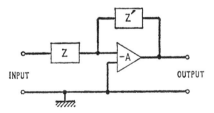

Fig. 1.6. Basic operational amplifier

The external voltage feedback loop across the amplifier through the impedance Z' combined with the series input impedance Z gives the complete operational amplifier certain useful properties. For example, where the internal amplifier gain is very large, it can be shown that the operational amplifier gain is approximately Z'/Z. The description 'operational' arises because this type of amplifier may be used to accomplish a number of mathematical operations.

With transistors, the basic operational amplifier takes the form of Fig. 1.7, where a common-emitter configuration provides the necessary real negative voltage gain.

16 *Elements of Transistor Pulse Circuits*

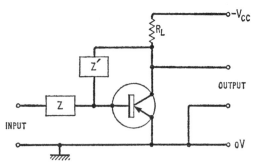

Fig. 1.7. Transistor operational amplifier

Sign changer

To change the sign of a signal (i.e. phase change 180° without amplitude change), Z' is made equal to Z. For d.c. signals this

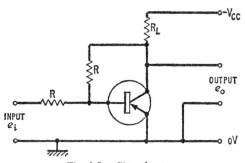

Fig. 1.8. Sign changer

would take the basic form shown in Fig. 1.8 where the feedback impedances Z and Z' are equal resistances R.

For a.c. signals, however, the two resistors R could be complex impedances, i.e. any combination of R, L and C, although usually, for simple sign change, resistors alone are used to ensure a phase shift independent of frequency.

Scale changer

To change the scale of a signal (i.e. amplitude change by a factor k), the feedback components of the operational amplifier should be selected to have $Z'/Z = k$, a real constant. Again, scale change is usually effected with Z and Z' selected as resistors, which gives the circuit of Fig. 1.9. In this, the output voltage is of opposite sign to

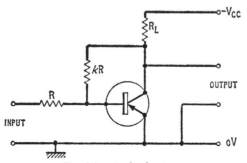

Fig. 1.9. *Scale changer*

the input, and increased by a factor k. Often the analogy is drawn here between the scale-changer operational amplifier and a lever with its fulcrum at the transistor base. The output volts go down as the input volts go up and the output volts travel is k times that at the input.

Phase shifter

To change only the phase of a sinusoidal a.c. signal, the series and feedback impedance components of the basic operational amplifier of Fig. 1.7 should be made equal in magnitude but should differ in phase angle. By using suitable values of capacitance or inductance, by themselves or with resistors, any phase shift from 0° to 360° may be obtained at any selected frequency.

Integrator

A specific example of phase shifting is the integrator circuit of Fig. 1.10 where the series input impedance is a resistor and the feedback impedance a capacitance. Here it can be shown that the output voltage e_o is related to the input voltage e_i by the formula

$$e_o = - \frac{1}{RC} \int e_i dt$$

The amplifier thus provides an output signal which is proportional to the time integral of the input voltage, i.e. it is an integrator.

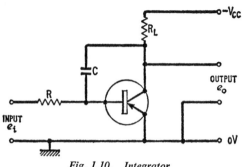

Fig. 1.10. Integrator

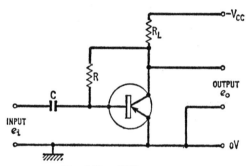

Fig. 1.11. Differentiator

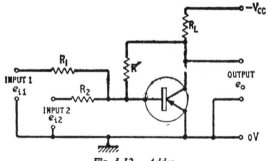

Fig. 1.12. Adder

Differentiator

Another phase-shift version of the operational amplifier is the differentiator circuit shown in Fig. 1.11. Here the series input impedance is a capacitor C and the feedback impedance a resistor R. It can be shown that the output voltage e_o is related to the input voltage e_i by the formula

$$e_o = - RC\, de_i/dt$$

The amplifier thus provides an output voltage which is proportional to the time derivative of the input voltage, i.e. it is a differentiating circuit or 'differentiator'.

Adder

A final use of the operational amplifier is to obtain a single output voltage which is a linear combination of a number of input circuits. This is illustrated in Fig. 1.12, with two inputs, where the input series impedances R_1 and R_2 and the feedback impedance R' are all resistive. It can be shown that the output voltage e_o in this case is related to the input voltages e_{i_1} and e_{i_2} as follows:

$$e_o = - \frac{R'}{R_1} e_{i_1} - \frac{R'}{R_2} e_{i_2}$$

Thus the output voltage is linearly related to the two input voltages.
If now R_1 is made equal to R_2, the output voltage is given by

$$e_o = - \frac{R'}{R_2} (e_{i_1} + e_{i_2})$$

This makes the output voltage proportional to the sum of the input voltages. The circuit is then known as an 'adder'. The addition action has been demonstrated for two inputs only but clearly more than this could be used.

In the more general case where the input resistors have different values, by suitable selection of resistors the scale of each input can be adjusted before adding.

This is only one of the many methods of combining a number of signals but it has the advantage that it may be extended to a very large number of inputs requiring only one additional resistor for each input. With a sufficiently high gain amplifier, there is a minimum of interaction between the input sources.

LINEAR SWEEP GENERATORS

A common requirement in non-linear circuitry is a linear sweep generator which produces an output voltage varying linearly with time. The simplest linear sweep is obtained by suddenly applying a direct voltage, V, to a resistor R and a capacitor in series as shown in Fig. 1.13 and taking the voltage across the capacitor as output. The resulting voltage v_o obeys the equation $v_o = V(1 - e^{-t/CR})$.

This gives a nearly linear rise in voltage so long as t is very much less than CR. Indeed it can be shown that the deviation from a linear rise $v_o = Vt/CR$ is less than 5% if t does not exceed CR/10,

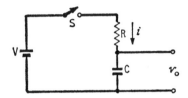

Fig. 1.13. Capacitor charge circuit

or if the output voltage does not rise above V/10. The rate of change of the output voltage can be shown to be given by

$$dv_o/dt = i/C$$

where i is the charging current through the resistance. The more constant i is, the better the linearity of the sweep.

One method of improving linearity is to use the constant-current collector characteristic of the transistor whose base input current is fixed. This is illustrated in Fig. 1.14 where the base current of the transistor Q is fixed (neglecting the small forward drop in the base emitter diode) by the base supply voltage V_{BB} and the base input series resistor R_B. When the switch S is closed, the voltage V is

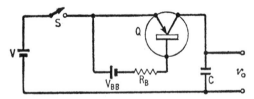

Fig. 1.14. Transistor linear capacitor charging

applied *via* the capacitor C between the collector and emitter of the transistor, which therefore conducts, the voltage polarities being correct for a p-n-p transistor as shown. As the transistor base current is fixed, so is its collector current. Thus the capacitor charging current is fixed and the output voltage rise is virtually linear. The charging current can be controlled by varying R_B and consequently the transistor collector current.

Bootstrap circuit

A better method for linear charging of a capacitor, however, is to use a large amount of negative feedback to keep the voltage across the

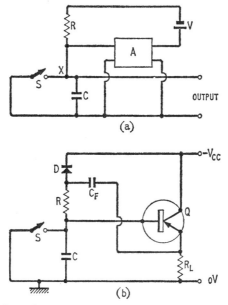

Fig. 1.15. (a) Basic bootstrap integrator; (b) Transistor bootstrap integrator

series charging resistor R of Fig. 1.13 constant. The feedback circuit for achieving a constant charging current into the capacitor is the well-known 'bootstrap' circuit. The basic arrangement is shown in Fig. 1.15 (a). When the input switch, S, is closed to short-circuit the capacitor C, the supply battery voltage V is effectively applied between earth and the top of resistor R, and C is discharged. If now switch S is opened, any voltage applied from V *via* R to the point X at the amplifier input, would reappear at the output of amplifier A (unity gain) and be applied *via* battery V to the top end of R. Thus the direct voltage across R (and thus the current through it) would remain constant although the voltage at X changes. Thus C charges up with a constant current and we get a linear voltage rise at the output. It should be noted that amplifier A not only should have unity gain, but must have a high input impedance so as to take negligible input current and low output impedance so as to have negligible voltage drop across the output. The bootstrap circuit is a true integrator, whose name arises from its ability to 'pull itself up by its own bootstraps'.

The basic transistor version of the bootstrap integrator is given in Fig. 1.15 (b). Here the switch S is normally closed and capacitor C is discharged. When S is opened, transistor Q acts as a unity gain amplifier to transfer the rising voltage on C to the top end of R. The capacitance C_F is very large compared with C and the voltage across it does not vary significantly as C charges up. Thus C_F substitutes for the battery V in Fig. 1.15 (a). The current through R is virtually constant and the voltage in C rises linearly. The emitter follower Q has the requisite unity gain, high impedance and low output impedance. The diode D is included so that as the voltage at the top end of R is bootstrapped up, and rises above the rail voltage, it reverse-biases the diode and cuts off the timing circuit from the power supply.

Miller integrator

Another method for improved linear charging up of a voltage across a condenser is the Miller 'integrator' shown schematically in Fig. 1.16 (a). Normally switch S is closed, and capacitor C discharged. When S is opened C begins to charge up through R, and it can be shown that due to the feedback through C, the output rate of rise is the same as would have been achieved with a capacitance $A \times C$ in series with a resistance R across a voltage supply $A \times V$, where A is the voltage amplification factor of the amplifier. This effective multiplication of the capacitance by the amplifier gain is of course the Miller effect and hence the description 'Miller integrator'. The apparent increase of both capacitance and rail voltage lead to a much more linear voltage rise than would have been achieved with the same values without the feedback amplifier.

A simple transistor version of the Miller integrator is given in Fig. 1.16 (b). Here the transistor Q acts as the high gain amplifier. Normally switch S is closed and transistor Q, with base then con-

Semiconductor and Pulse Circuits 23

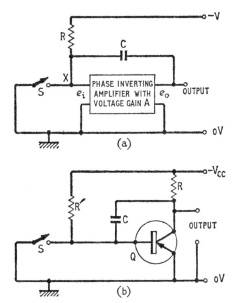

Fig. 1.16. (a) Basic Miller integrator; (b) Transistor Miller integrator

nected to earth is cut off. With no collector current flowing, there is no voltage drop in R, the output is at rail voltage and the capacitor C is fully charged. When S is opened, base current begins to be supplied through R' and the transistor begins to switch on. By the feedback through C as explained earlier, the voltage at the output falls linearly from the rail voltage level towards earth.

HIGH INPUT-IMPEDANCE AMPLIFIER

Darlington pair

In pulse circuitry there is often a requirement for an amplifier with a high input resistance and a low output resistance. The emitter follower discussed earlier does provide these to some extent, but where they are required to a higher degree, use is often made of the compound emitter follower or 'Darlington pair' illustrated in Fig. 1.17. To a first approximation, the current gain of this circuit is equal to the product of the current gains of the individual transistors. The input resistance is equal to the emitter load resistance multiplied by this current gain product and the output resistance to the source

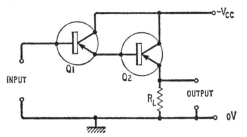

Fig. 1.17. *Darlington pair*

resistance divided by this product. From this it is clear that very high input and very low output resistances are possible.

Bootstrap amplifier

Another circuit used to achieve high input resistance, especially where base-bias resistor networks are likely to shunt the signal input is the bootstrap input circuit in Fig. 1.18. Here negative feedback introduced by the unbypassed emitter resistor gives a high input impedance at the base of the transistor Q. To prevent this being shunted significantly by the base bias network R_1, R_2, the base bias current is supplied through an isolating resistor R_3 from the centre point of R_1 and R_2. A capacitor C connected between the transistor

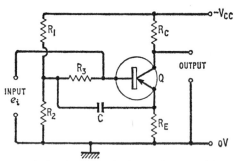

Fig. 1.18. *Bootstrap high-impedance amplifier*

emitter and the bottom end of R_3 feeds back a signal voltage almost equal to the voltage at the top end of the emitter resistor R_E. It can be shown (if C is large enough for its reactance to be neglected at the frequency considered) that the signal voltage across R_3 is $(e_i - Ae_i)$, where A is the voltage amplification of the transistor. Thus the bias network draws from the signal source a current of only $e_1 (1 - A)/R_3$.

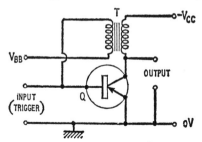

Fig. 1.19. Blocking oscillator

This means in effect that R_3 is effectively multiplied by $1/(1 - A)$ so far as the a.c. signal current is concerned, while still representing its own value for d.c. bias purposes. With high gain transistors, A can be brought very close to unity so that $1/(1 - A)$ can be very large, and the bias network does not significantly shunt the signal input. This is only one further example of 'bootstrapping', i.e. multiplying the apparent value of a resistor by applying nearly equal in-phase signal voltages at each end.

REGENERATIVE SWITCHING CIRCUITS

The circuits surveyed so far have all been non-regenerative. No survey of pulse circuits would be complete without mention of the group of important switching circuits which use large positive feedback to give regenerative switching between two discrete states, i.e., 'two-state' circuits.

Blocking oscillators

One of the most ubiquitous regenerative switching circuits is the blocking oscillator. This is essentially a transformer-coupled oscillator, with regenerative feedback from output to input large enough to cause the transistor to become either saturated or cut off over a substantial part of the operating cycle. Fig. 1.19 illustrates the basic circuit for transistors in common-emitter form. Feedback is obtained by the phase-reversing transformer T. Output can be taken directly from the collector as shown or from a tertiary winding on the transformer. It is usually arranged to produce output pulses of large magnitude and short duration. By suitable choice of the bias V_{BB}, the circuit can be made to give a periodic train of pulses (astable or free running), or single pulses when triggered by a suitable pulse input.

The blocking oscillator is often used as a clock oscillator to generate a continuous train of accurately-controlled short pulses to synchronise a series of switching operations. It is also used to obtain abrupt pulses from a slowly-varying input triggering voltage, i.e., for pulse reshaping. It can be easily arranged to generate pulses of very large peak power with a low duty cycle. It also finds uses as a frequency divider, a low-impedance switch and a gating voltage source.

Multivibrators

Another regenerative two-stage circuit frequency met with is the multivibrator. In its commonest transistor form, as in Fig. 1.20 (a) this is a two-stage common-emitter amplifier with phase inversion over each stage, and with the output heavily coupled back to the input. This heavy positive feedback gives the circuit the property of switching rapidly between two extreme states, with one transistor switched hard on and the other cut-off. How long it stays in the extreme states is decided by the cross coupling impedances Z and Z′ selected. There are three basic possibilities all derived from the general circuit shown in Fig. 1.20 (a).

Astable multivibrators

When both cross-coupling impedances Z, Z′ are capacitances, we get the circuit of Fig. 1.20 (b), which is free running (astable), and produces at each collector a train of rectangular pulses, without external triggering. Hence it is often called a square wave generator.

Bistable multivibrator

When both cross-coupling impedances of Fig. 1.20 (a) are resistances we have the bistable multivibrator illustrated in Fig. 1.20 (c). This can exist indefinitely in either of two stable states with one transistor on and one transistor off, but also it can be caused to make an abrupt transition from one state to the other by a suitable external trigger pulse. It finds extensive application in pulse circuitry to generate square waves from pulses, and for certain digital operations such as counting.

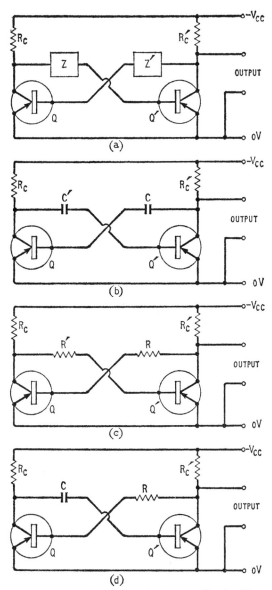

Fig. 1.20. (a) Basic multivibrator, (b) Astable multivibrator, (c) Bistable multivibrator, (d) Monostable multivibrator

Monostable multivibrator

The last member of the multivibrator branch of the pulse circuit family is the monostable multivibrator shown in Fig. 1.20 (d). The use of a resistor and a capacitor for cross coupling in this circuit gives the curious property of one permanently stable state (Q on and Q' off)

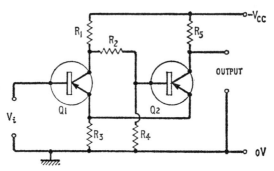

Fig. 1.21. *Schmitt trigger*

and one quasi-stable state (Q off and Q' on). The circuit normally lies in its stable state but if suitably triggered it passes abruptly into the quasi-stable state for a time which is long compared with the transition time between states. Eventually, however, this multivibrator will return abruptly to its stable state on its own without any external signal being required to produce the reverse transition.

The primary use for the monostable multivibrator is to establish a time interval which starts on the application of a pulse but whose length is independent of the length of the trigger pulse. It is much used for pulse reshaping and for establishing preset delays.

Schmitt trigger

There are many derivatives of the blocking oscillator and multivibrator circuits described above, but in this brief preliminary survey mention can be made of only one. This is the Schmitt trigger, illustrated basically in Fig. 1.21. It has the interesting property that when the input voltage V_i is below a certain level, called the triggering level, the transistor Q2 is hard on and Q1 cut off. When V_i rises above the triggering level, there is an abrupt changeover, Q1 switching on and Q2 off. If V_i is reduced again below the trigger level, the circuit returns to its original state. From that it is clear that the Schmitt trigger is neither bistable nor monostable in the ordinary sense. It behaves like a non-regenerative switch controlled by the input d.c. level, but has the advantage that it switched abruptly at very high speed and can be designed to have an accurate adjustable trigger level. The circuit is very widely used to produce a square wave from a slowly varying input of irregular pulse shape, and as a sensitive voltage level detector.

CONCLUSION

This preliminary brief survey of transistor non-linear circuits is necessarily incomplete, but is designed to give the reader some idea of the variety of such circuits in common use, and how they differ from the linear small-signal circuits which are the main building blocks in the communications field. Later the properties and applications of these non-linear circuits will be dealt with in more precise detail.

CHAPTER 2

Linear Pulse Amplifiers

In pulse work, an obvious need is for a circuit to amplify a pulse with as little change of shape as possible, i.e. a linear pulse amplifier. As it happens, this also turns out to be what most engineers know as either a video, wide-band, broad-band or base-band amplifier. What you call it depends on what field you work in, but it is basically an amplifier whose passband extends from a low audio frequency to an upper frequency many times the lower limit frequency. The high frequency limit is usually not less than 100 kHz. It may in special cases extend out to many hundreds of megahertz.

Transistors are smaller and more efficient than valves and have tended to replace them in general-purpose linear pulse amplifiers. Fortunately, most of the wide-band amplifier techniques developed for valves carry over into transistors. We will deal only briefly with these established techniques, leaving it to the interested reader to consult a standard text-book such as *Pulse and Digital Circuits*, by J. Millman and H. Taub (McGraw-Hill). Special problems raised by the use of transistors we will examine more comprehensively.

GENERAL DESIGN PROBLEMS

There are two ways you can approach linear pulse amplifier design. The first is to consider the response of the amplifier to a square-wave input and specify the performance in terms of (a) pulse gain, (b) pulse rise-time, and (c) pulse droop. The second is to examine its response to a continuous sine-wave input, and specify performance by (a) midband gain, (b) upper and (c) lower half-power bandwidth limits. It can be shown that the two approaches lead to the same practical results and we will generally adopt the more familiar continuous sine-wave approach.

Interstage coupling

Interstage coupling is the first problem. In theory, any of the three conventional methods—RC, direct or transformer coupling—is possible. In practice, linear pulse amplifiers are almost always RC-coupled. Direct or d.c. coupling brings in such problems of drift and of setting up bias voltage levels, that it is used only where the amplifier is specially required to operate down to zero or near-zero frequency.

Transformer coupling, too, raises difficult problems, mostly in the design of the transformer itself. With ferrite cores, a 30 Hz–5 MHz bandwidth is feasible. Bandwidths up to 200 MHz have been produced experimentally with transformers, but, with these, optimum interstage matching is not usually possible due to the limited number of wire turns. Recent developments in wide-band transformer-coupled amplifiers employing transmission-line-type winding techniques have made possible bandwidths out to 1 GHz at the high-frequency end, but poor low-frequency response restricts their use in linear pulse amplifiers. Quite apart from the design difficulties, transformer coupling is relatively the most expensive coupling method.

Because of all this, we will confine ourselves to the design of RC-coupled pulse amplifiers only.

Transistors in linear pulse amplifiers

The transistor is normally considered as a current, rather than a voltage, amplifier, and the design of transistor linear pulse amplifiers usually works in terms of current gain. This contrasts with valve designs which are usually worked in terms of voltage gain.

In a linear pulse amplifier, the transistor could be used in any of the three basic configurations shown in Fig. 2.1: (a) common-base; (b) common-emitter, and (c) common-collector. However, the common-emitter connection is normally used, because with RC coupling it can provide both voltage and current gain. Without a matching transformer the common-base arrangement has no current gain and the common-collector no voltage gain. The common-collector and common-base connections will mainly be found used

32   *Elements of Transistor Pulse Circuits*

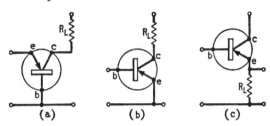

Fig. 2.1. Transistor configurations: (a) *common-base,* (b) *common-emitter,* (c) *common-collector*

only as terminal stages. The common-collector may start or end a chain to give high input resistance or low output resistance. The common-base may be used as a final stage where high voltage is required.

Pulse characteristics

As noted earlier, one way of specifying the small-signal performance of a linear pulse amplifier is in terms of pulse properties such as rise and fall times. Fig. 2.2 illustrates the terms used for a typical pulse waveform. AXYS represents the input pulse, and ABCDEFGH the amplified output pulse. The *pulse gain* is the ratio of RD, the output pulse peak amplitude, to AX, the input amplitude. The *rise time*, T_R, is the distance PQ along the time axis where PB represents 10% and QC 90% of the peak amplitude, RD, i.e. the time for the output pulse to rise from 10% to 90% of the peak. At the end of the input pulse, S, the output pulse will have 'drooped' to the point E, and the *pulse droop* is the fall JE expressed as a percentage of the peak amplitude, RD. Finally the *fall time*, T_F, is the distance TU along the time axis corresponding to the fall from point F (10% below point E) to point G (90% below point E). In the diagram, we have also shown in a 'dotted' curve, CZD, the 'overshoot' that sometimes occurs when inductive compensation circuits are used. Here the vertical distance from D to Z, expressed as a percentage of RD, is a measure of the overshoot.

As we are considering small-signal operation, the transistor operates only in its active region, neither cutting off nor bottoming. The droop of the pulse top then depends on the relation of the pulse length to the time constants of the coupling and by-pass RC networks. Also, provided the droop is small (as is usually the case), the fall time can be taken as equal to the rise time.

Linear Pulse Amplifiers 33

The pulse performance is then normally specified by (a) gain, (b) rise time, and (c) percentage droop. (For the present we are disregarding overshoot.)

It can be shown that to a good approximation, the response to a train of square waves is related to continuous response characteristics as follows:

$$\text{Pulse gain} = \text{midband gain} \quad (2.1)$$
$$\text{Pulse rise (and fall) time} = 1/3f_B \quad (2.2)$$
$$\text{Percentage pulse droop} = 300f_o/f \quad (2.3)$$

where

f_B = amplifier continuous response upper 3 dB turnover frequency,

f_o = lower 3 dB turnover frequency, and

f = repetition frequency of input square wave.

Frequency characteristics

In a linear pulse amplifier, the primary circuit requirement is to keep the gain constant over the amplifier design bandwidth. For various reasons, apparent later, gain tends to fall off at both low and high frequencies.

At low frequencies, the transistors themselves present few problems. Their parameters are real and constant in this region. Input and output resistances can be easily calculated from simple formulae and conventional audio-frequency design techniques used. Low-frequency response falls off mostly due to the limitations of finite coupling and bypass capacitors. The main difference from

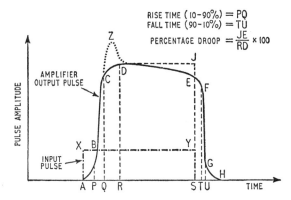

Fig. 2.2. Pulse response characteristics

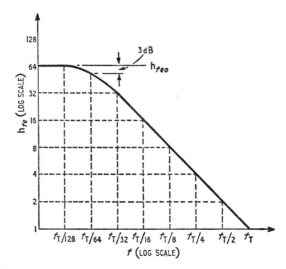

Fig. 2.3. Variations of common-emitter current gain (h_{fe}) with frequency

valve designs will be found in the values of the coupling capacitors. Transistor input impedance being low compared with a valve, coupling capacitors generally come out orders of magnitude higher. The actual physical size of the capacitors does not increase, however, because the transistor works at a much lower voltage level than a valve and capacitors with lower voltage ratings can be used.

At high frequencies, amplifier response falls off mainly due to the fall off in transistor gain, and to transistor internal capacitances or circuit external stray capacitances shunting the signal or introducing feedback.

In the transistor itself, the common-emitter short-circuit current gain falls off to unity at a characteristic frequency f_1 (practically the same as the frequency f_T sometimes quoted instead by manufacturers). This behaviour is illustrated in Fig. 2.3. It will be seen that above f_{hfe} (the frequency at which h_{fe} is 3 dB down on its low frequency value, h_{feo}) the transistor current gain follows the approximate law:

$$h_{fe} \text{ (at frequency } f) = f_T/f \qquad (2.4)$$

and that f_{hfe} is therefore given by

$$f_{hfe} = f_T/h_{feo} \qquad (2.5)$$

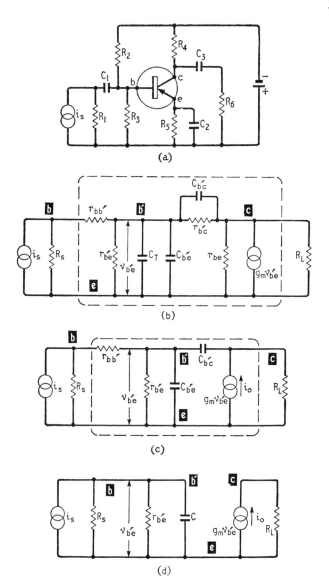

Fig. 2.4. Derivation of simplified equivalent circuit for high-frequency linear pulse amplifier, showing (a) actual common-emitter amplifier, (b) full hybrid-pi equivalent circuit, (c) partially-simplified equivalent circuit and (d) fully-simplified equivalent circuit

36 *Elements of Transistor Pulse Circuits*

In this context, we should note a concept often used in amplifier design—the 'gain-bandwidth' product, GB. This is the product of the midband gain and the 3 dB upper frequency limit. For the transistor itself,

$$\text{GB} = h_{feo} \times f_{hfe} = h_{feo} \times (f_T/h_{feo}) = f_T \quad (2.6)$$

This gain-bandwidth product, f_T, is often used as a figure of merit for a transistor to be used in a linear pulse amplifier. Clearly, the higher f_T, the less the current gain fall-off will restrict the upper frequency response of the amplifier.

Equivalent circuits

There are several equivalent circuit models that could be used for considering transistor high-frequency performance in a linear pulse amplifier. At the approach level adopted here, the simplified hybrid-pi circuit developed in Fig. 2.4 has some advantages. Fig. 2.4 (a) shows a simple RC-coupled common-emitter stage. At the high-frequency end of the bandwidth we can consider the coupling and bypass capacitors, C_1, C_2, C_3, as a.c. short circuits. Also we can lump the source resistance, R_1, and the two base-bias resistors, R_2, R_3, into one equivalent resistance, R_s. Finally we can lump the collector resistor, R_4, and the load resistor, R_6, into an equivalent resistance, R_L. Using the full hybrid-pi equivalent circuit, this leads to the arrangement of Fig. 2.4 (b), where each of the elements is independent of frequency.

In linear pulse amplifiers, R_L is usually small compared with $r_{b'c}$ and r_{ce}, and C_T is relatively small compared with $C_{b'e}$. Ignoring these high-impedance items, we get the simplified hybrid-pi circuit of Fig. 2.4 (c). As the load resistance is usually small, $C_{b'c}$ does not shunt it appreciably. Moreover, the direct transmission through $C_{b'c}$ is small compared with the output of the current generator $g_m v_{b'e}$, and may be ignored on the output side. Because of the Miller effect, the effective capacitance looking into $C_{b'c}$ towards the load can be shown to be approximately $g_m R_L C_{b'c}$ (which is effectively in parallel with $C_{b'e}$). Finally, in practical circuits, $r_{bb'}$ is usually small compared with R_s and $r_{b'e}$, so that the equivalent circuit can be further simplified to Fig. 2.4 (d), where $C = C_{b'e} + g_m R_L C_{b'c}$.

Looking at the equivalent circuit of Fig. 2.4 (d), we can see that the voltage, $v_{b'e}$, across $r_{b'e}$ will fall off at high frequencies due to the

Linear Pulse Amplifiers 37

shunting effect of C. This means that the corresponding amplifier output current, $g_m v_{b'e}$, also falls off at high frequency.

Amplifier gain and bandwidth without compensation

For the simplified circuit of Fig. 2.4 (d), which has no frequency compensation circuit included, the current gain is given by

$$A_i = \frac{i_o}{i_s} = \frac{g_m v_{b'e}}{v_{b'e}/(R_s // r_{b'e} // X_c)}$$

$$= g_m(R_s // r_{b'e} // X_c) \qquad (2.7)$$

At low (midband) frequencies, X_c, the reactance of C, can be disregarded as very high, and the midband amplifier current gain is

$$A_{io} = g_m(R_s // r_{b'e}) = g_m \times \frac{R_s r_{b'c}}{R_s + r_{b'e}} \qquad (2.8)$$

As frequency rises, the current gain A_i in equation (2.7) falls to a 3 dB-down point at a frequency given by

$$f_B = \frac{1}{2\pi C(R_s // r_{b'e})} = \frac{R_s + r_{b'e}}{2\pi C R_s r_{b'e}} \qquad (2.9)$$

From (2.8) and (2.9), the amplifier gain-bandwidth product is

$$GB = A_{io} f_B = \frac{g_m}{2\pi C} = \frac{g_m}{2\pi(C_{b'e} + g_m R_L C_{b'c})} \qquad (2.10)$$

If in Fig. 2.4 (d) we make R_s very large and R_L very small, we get the circuit for measuring the short-circuit current gain of the transistor itself. From equation (2.8) this gives

$$h_{feo} = g_m r_{b'e} \qquad (2.11)$$

and, from (2.9)

$$f_{hfe} = \frac{1}{2\pi C_{b'e} r_{b'e}} \qquad (2.12)$$

Thus the transistor gain-bandwidth product referred to in (2.6) is

$$f_T = h_{feo} \times f_{hfe} = \frac{g_m}{2\pi C_{b'e}} \qquad (2.13)$$

38 *Elements of Transistor Pulse Circuits*

Typical uncompensated linear pulse amplifier stage

To put real values to all this, let us consider a modern germanium alloy r.f. transistor, for which typically, $g_m = 39 \times 10^{-3}$ mho, $r_{b'e} = 2,500$ ohms, $C_{b'e} = 530$ pF, and $C_{b'e} = 10$ pF. Substituting these values in (2.11), (2.12) and (2.13) we get $h_{feo} = 100$, $f_{hfe} = 125$ kHz, and $f_T = 12 \cdot 5$ MHz. (Note that in alloy r.f. transistors it can be shown that $f_{co} = 1 \cdot 2 f_T$, and we can thus derive the more familiar $f_{co} = 15$ MHz).

If we now take a typical source resistance $R_s = 1,000$ ohms and a typical load resistance $R_L = 800$ ohms, and substitute these values in (2.8), (2.9) and (2.10), we get achievable characteristics of an actual single-stage uncompensated linear pulse amplifier as follows: Midband current gain = 28, 3 dB bandwidth = 280 kHz and gain—bandwidth product = 7·8 MHz. In terms of pulse response, this gives from (2.1) and (2.2) earlier, a peak pulse gain of 28 and a pulse rise (and fall) time of 1·2 μsec.

Collector capacitance effect

In all the above, we have ignored the possible shunting of R_L by the collector capacitance. That this is justified we can see as follows: It can be shown that the collector-emitter capacitance is approximately given by $A_i \times C_{b'c}$. This shunts the load R_L, and the half-power frequency cut off of the combination in the numerical example given is:

$$f_c = \frac{1}{2\pi R_L(A_i C_{b'c})} = 710 \text{ kc/s} \qquad (2.14)$$

which is well beyond the 280 kHz bandwidth of the amplifier arrived at above, when we ignored the shunting of R_L.

HIGH-FREQUENCY COMPENSATION

Four main techniques are used to extend the high-frequency response of linear pulse amplifiers: (1) negative feedback, (2) shunt inductance peaking, (3) series inductance peaking and (4) positive feedback.

Negative feedback H.-f. compensation

Simple negative feedback can be used in two ways as shown in Fig. 2.5. In the first case, Fig. 2.5 (a), the unbypassed emitter

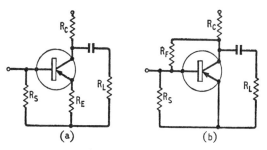

Fig. 2.5. *High-frequency compensation by resistive negative feedback using (a) unbypassed emitter resistor and (b) collector-base feedback resistor*

resistor R_E, lowers the mid-band gain and pushes out the bandwidth limit, because the gain-bandwidth product is not much affected. It can be shown that the method is effective for R_E in the range:

$$R_s/h_{feo} < R_E < R_s \qquad (2.15)$$

Approximate formulae for the compensated amplifier current gain and bandwidth are then:

$$A_{io(F)} = \frac{A_{io}}{1 + kh_{feo}} \qquad (2.16)$$

and

$$f_{B(F)} = f_B \left(1 + \frac{kh_{feo}}{1+k}\right) \qquad (2.17)$$

where $k = R_E/R_s$.

In the typical numerical example given earlier, we had $A_{io} = 28$, $f_B = 280$ kHz, $R_s = 1{,}000$ ohms and $h_{feo} = 100$. If we now take $k = 1/100$, this gives $R_E = 10$ ohms, and we find that with feedback A_{io} drops to 14 and f_B widens to 560 kHz.

A second method of applying simple negative feedback for gain-bandwidth trading is shown in Fig. 2.5 (b) where feedback is from collector to base via the resistor R_F. It can be shown that this method is effective for R_F in the range.

$$R_L < R_F < h_{feo}R_L \qquad (2.18)$$

However the technique is of limited application for very wide bandwidths because the feedback resistor R_F eventually becomes

40 *Elements of Transistor Pulse Circuits*

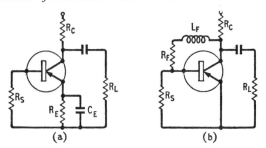

Fig. 2.6. High-frequency peaking by selective negative feedback, derived by (a) shunting emitter resistor by small capacitor and (b) inserting small inductance in series with feedback resistor

so low that it materially loads both input and output circuits with a consequent loss of gain-bandwidth product.

The two simple negative feedback circuits of Fig. 2.5, can be modified for high-frequency peaking by introducing frequency conscious elements in the feedback path as shown in Fig. 2.6. Shunting of the emitter resistor by a small capacitor C_E as in Fig. 2.6 (a) reduces the feedback at high frequency and thus increases the amplifier high-frequency gain. The value of C_E is often selected initially so that its reactance at f_{hfe} is equal to R_E, i.e.

$$C_E = 1/(2\pi f_{hfe} R_E) \qquad (2.19)$$

With this as a starting value, C_E can be adjusted experimentally for the desired results.

The simple collector-base feedback of Fig. 2.5 (b) can also be modified for high-frequency peaking by padding out the feedback resistor R_F with a series inductor L_F as shown in Fig. 2.6 (b). This inductor tends to reduce the effect of the feedback resistor R_F at high frequencies and thus extends the bandwidth. It can be shown that an approximate value for this inductor is:

$$L_F = R_F \frac{(1 + R_F/R_L)}{2\pi f_T} \qquad (2.20)$$

Shunt-inductance H.-f. compensation

Shunt-inductance peaking can take several forms as illustrated in Fig. 2.7. The basic arrangement is an inductive impedance in parallel with the load, R_L, as in Fig. 2.7 (a). Signal current is drawn off from the load through R and L in series. At low frequencies, the reactance X_L of L is low and the current diverted from the load is $R_L/(R + R_L)$ of the total available. At high frequencies, X_L increases and the current diverted from the load is reduced; i.e. the current into the load increases, compensating for the fall-off occurring otherwise.

In practical circuits, the shunt compensation is often achieved by padding out the collector resistor, R_c as shown in Fig. 2.7 (b). Although the value of L is usually set by 'cut and try' methods, a useful starting point is to select L so that its reactance at the transistor common-emitter cut-off frequency, f_{hfe}, is equal to R_c, i.e.

$$L = R_c/(2\pi f_{hfe}) \qquad (2.21)$$

The value of L to be expected in this case for the typical alloy r.f. transistor described earlier with $f_T = 12\cdot 5$ MHz, $h_{feo} = 100$ and $R_c = 1,500$ ohms is then about 2 mH.

The compensating inductor may be inserted in some other parallel network such as the base-bias potentiometer as illustrated in Fig. 2.7 (c). Here the same starting rule of thumb—making the reactance of L at f_{hfe} equal to R_B—can be used.

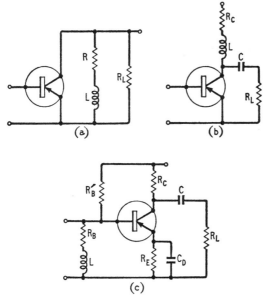

Fig. 2.7. *Shunt high-frequency peaking circuits. At (a) is shown load-resistor padding, (b) padding of collector-load resistor and at (c) base-bias-resistor padding*

42   *Elements of Transistor Pulse Circuits*

Shunt peaking is useful for compensating the current-gain fall-off of the transistor, but another method, series peaking, is usually adopted for dealing with signal shunting by the transistor output capacitance.

Series-inductance H.-f. compensation

Series inductance peaking is illustrated in its basic form in Fig. 2.8, and consists of inserting an inductor L in series with the transistor output. This inductor is approximately chosen initially to resonate with the collector output capacitance, $A_{to} \times C_{b'c}$, at the collector cut-off frequency:

$$f_C = \frac{1}{2\pi A_{to} R_L C_{b'c}} \tag{2.22}$$

It is then adjusted by cut and try methods. The practical circuit takes the form of Fig. 2.8 (b) where the effective load resistance is R_c and R_o in parallel, and the large coupling capacitor C has such a low reactance that it can be ignored in the calculations. The optimum inductance value is difficult to compute because of the many variables involved, but in practice, designers often start from the empirical assumption that its reactance at the cut-off frequency of the unpeaked amplifier is equal to the transistor short-circuit input impedance at that frequency (usually obtainable from the manufacturer's data sheet).

Other H.-f. compensation techniques

Positive feedback at high frequency is sometimes introduced by design to push out the upper frequency limit of the amplifier. This should be done with caution, because spreads of transistor characteristics can easily lead to instability.

Finally, it can be shown that the gain-bandwidth product of an uncompensated transistor wide-band amplifier is relatively independent of the emitter bias current, I_e. However, the transistor gain varies directly with I_e, so that it is possible to trade gain and bandwidth by varying I_e. This method has the disadvantage that the gain-bandwidth can vary widely between units because of the spread of transistor gain.

We have been concentrating on the transistor aspects of high-frequency compensation, but the reader is reminded that the valve technique of gain-bandwidth trading by varying interstage resistance loading is applicable also to transistors. The only limitation is that the resultant interstage shunt resistance R_s must lie between $r_{bb'}$ and $r_{b'e}$.

Another point not to be forgotten is that the overall amplifier gain must remain stable, i.e. not break into oscillation. Rigorous theoretical examination of possible instability can be made, but is very difficult, and, transistor characteristic spreads being so wide, this aspect is usually covered by experimental testing.

A practical aspect of high-frequency compensation is the method of testing bandwidth characteristics. We have been thinking in

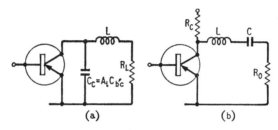

Fig. 2.8. Series-inductance high-frequency peaking, showing the basic circuit (a) and at (b) the practical circuit

terms of conventional sine-wave response measurements, but, with modern fast oscilloscopes available, it is often more convenient to inspect and measure the pulse response visually. This is where the 'overshoot' mentioned earlier becomes significant, because the appearance of an overshoot on the output pulse leading edge gives an indication of impending instability that might not be evident from a continuous-signal bandwidth measurement. Usually a few per cent of overshoot is allowable to sharpen the leading edge of the output pulse before instability becomes a significant problem.

The analysis of high frequency compensation techniques given has been very approximate, ignoring many second order effects and making severe practical simplifying assumptions. However, the results will be found correct to a first order of magnitude. In general, it will be found that by judicious use of peaking techniques, bandwidths can be approximately doubled for any specified gain. Anyone seeking a more rigorous treatment should consult standard textbooks such as *Transistors—Principles, Design and Applications* by W. G. Gartner, Van Nostrand, 1960, or *Transistor Circuit Analysis* by M. V. Joyce and K. K. Clarke, Addison-Wesley 1961.

LOW-FREQUENCY COMPENSATION

The low-frequency cut-off (or percentage droop—see (2.3)) of a linear pulse amplifier is not usually much dependent upon the characteristics of the transistors; it is determined by the same considerations as apply to extending the low-frequency bandpass limit of a conventional RC-coupled audio amplifier. The main limiting

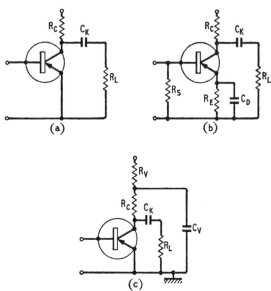

Fig. 2.9. Low-frequency response considerations. C_K in (a) and C_D in (b) both cause a fall-off at low frequencies. Circuit at (c) is one method of correction

factors are the inter-stage coupling capacitor, C_K, shown in Fig. 2.9 (a), and the emitter decoupling capacitor, C_D, in Fig. 2.9 (b). The design values of C_K and C_D are usually chosen to give not more than 1 dB attenuation each at the lower bandwidth limit, f_o, aimed at, so that in combination they give less than 3 dB attenuation. A formula for the interstage coupling capacitor on this basis is:

$$C_K = \frac{3}{2\pi f_o (R_C + R_L)} \qquad (2.23)$$

and for the emitter bypass capacitor

$$C_D = \frac{3 h_{feo}}{2\pi f_o R_s} \qquad (2.24)$$

Linear Pulse Amplifiers 45

Another method of low-frequency compensation well known from valve practice can also be used with transistors as shown in Fig. 2.9 (c). Here R_C and C_V in parallel give bass boost, provided R_V is chosen as high as possible consistent with the supply voltage, and a value of C_V is selected:

$$C_V = R_L C_K / R_C \tag{2.25}$$

TRANSISTORS FOR LINEAR PULSE AMPLIFIERS

When we use a transistor with a gain-bandwidth product, f_T, in a practical linear pulse amplifier stage, we generally find that due to various unavoidable losses in bias resistors, etc., it is not easy to attain an amplifier gain-bandwidth much more than 2/3 of f_T. Now transistors used in linear pulse amplifier fall into two main types: alloy and diffused base. The best commercially-available alloy types give a guaranteed minimum gain-bandwidth, f_{Tmin}, of about 6 MHz, and these can be used to produce a stage with a gain-bandwidth of about 4 MHz in practice, e.g. to produce a gain of 10 and a bandwidth of 400 kHz.

For higher gain-bandwidth requirements than this, recourse has to be made to diffused-base transistors, where a minimum f_T of 60 MHz is now obtainable almost as cheaply as the 6 MHz alloy type. Although the analysis given earlier applied primarily to alloy r.f. types, it still holds substantially true for diffused types, so that gain-bandwidths of some 40 MHz are easily obtainable in practice. In some of the ordinary linear pulse requirements, indeed, it may be found necessary to add circuit elements to restrict rather than to enlarge the amplifier band width. Some diffused types with f_{Tmin} of 300 MHz are now commercially available so that amplifiers with gain bandwidth products of 200 per stage are feasible.

MULTISTAGE LINEAR PULSE AMPLIFIER

Where the desired amplifier gain cannot be obtained with one stage, it is possible to cascade linear pulse amplifier stages effectively. The gain of a single stage of the type discussed in detail above is given by

$$A_i = \frac{A_{io}}{1 + jf/f_B} \tag{2.26}$$

46   *Elements of Transistor Pulse Circuits*

The gain of n identical stages would then be

$$A_{in} = A_i{}^n \qquad (2.27)$$

It can be shown that the overall bandwidth of n such stages is given by:

$$F_B = f_B/(1 \cdot 2 n^{\frac{1}{2}}) \qquad (2.28)$$

if n is greater than 3. Thus we can see that the overall bandwidth shrinks as the square root of the number of stages. It will be found also that as n increases the overall gain-bandwidth increases at first, but finally falls off. This means that a point is ultimately reached where adding an extra stage merely reduces the total gain-bandwidth product.

Readers looking for more information on multi-stage linear pulse amplifiers should consult *Handbook of Semiconductor Electronics* by L. P. Hunter, McGraw-Hill, 1962.

HIGH-LEVEL LINEAR PULSE AMPLIFIERS

Linear pulse amplifiers are often required to furnish large peak-to-peak voltage swings with bandwidths of many MHz for such applications as driving the grid or deflection plates of a cathode ray tube. Maximum output may be limited by the transistor voltage rating or by the useful range of current over which the transistor retains its high frequency performance. For a discussion of the practical problems involved the reader should consult *Transistor Television Receivers* by T. D. Towers, Iliffe Books Ltd., 1963.

In high-voltage linear pulse amplifiers with very wide bandwidths (over 25 MHz), unavoidable wiring capacitances and transistor output capacitances forces the use of small resistance loads. The maximum output will then be limited by the maximum permissible current swing (or in other words, the maximum dissipation). Where circuit strays, and not C_c, are the major frequency limiting factor, higher peak-to-peak load current can be obtained by paralelling transistors.

For frequencies less than 5 MHz, the voltage rating of the transistor tends to be the main limiting factor at present. In this case the use of the common-base configuration is advantageous, since common-base usually permits higher collector current excursions before distortion or breakdown. Common-base has the further advantage that wide alpha bandwidth permits the collector voltage

Linear Pulse Amplifiers 47

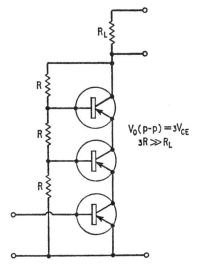

Fig. 2.10. Stacking transistors to give output swing greater than voltage ratings of individual transistors

to swing closer to zero without distortion than is the case with common-emitter. Typical practice is to drive the common-base output stage from a high output-impedance common-emitter stage.

If a voltage swing greater than the transistor rated voltage is desired, it is possible to connect transistors in a stack employing signal feedback to ensure all transistors share the output voltage equally. Such a configuration is shown in Fig. 2.10.

CONCLUSION

Modern transistors with gain-bandwidth products up to hundreds of megacycles make possible the use of transistors in most linear pulse amplifiers that are commonly required. Special cases requiring gain-bandwidths beyond this may require recourse to more exotic circuits such as transistor distributed amplifiers.

CHAPTER 3

Astable Multivibrators

In pulse electronics, a common requirement is for a circuit to give abrupt transitions between two different electrical states (usually high and low voltage). There are three main types of such two-state circuits:

(a) *astable*, i.e. switching continuously between the two states at a constant repetition rate without external excitation.
(b) *monostable*, i.e. normally lying in one state, but, when suitably triggered by an external signal, passing into the other, remaining there for a pre-set time, and then returning to the first state of its own accord without further triggering.
(c) *bistable*, i.e. capable of remaining indefinitely in either of two states, passing from one to the other only when externally triggered.

The astable two-state circuit is often referred to as 'free-running', and the monostable and bistable circuits as 'triggered'. This may perhaps be found a little ambiguous, because the astable circuit can also be triggered by a series of synchronising pulses of a higher frequency than its natural one.

THE MULTIVIBRATOR FAMILY

The most widely used two-state circuit is probably the 'multivibrator'. This is a two-stage amplifier with phase inversion over each stage and with the output heavily coupled back to the input. The feed-back is thus large and positive. Due to regeneration the circuit can then be made to pass very rapidly from one to the other of its two possible limiting states.

The multivibrator first made its appearance (with valves) in April 1918, in a report by H. Abraham and E. Bloch, *Notice Sur Les Lampes-Valves à 3 Electrodes et Leurs Applications* (Publication No. 27 of the French Ministère de la Guerre). In this form the multivibrator was free running, but W. H. Eccles and F. W. Jordan soon reported a bistable version (in the *Radio Review*, 1919, Vol. 1, page 143).

In one form or another, the multivibrator has proved to be a ubiquitous tool in pulse circuitry. Over the half century since its discovery it has been much analysed and discussed. It has been used under many different names. The convenient colloquial abbreviation 'multi' has become firmly accepted and we will use this for its brevity. Other terms that will be found in the literature are listed below.

(a) *Astable:* astable multi, multi (by itself), Abraham-Bloch circuit, free-running multi, square-wave generator, unstable multi, AMV.
(b) *Monostable:* monostable multi, monostable (by itself), univibrator, monovibrator, delay multi, one-shot, single-cycle, single-step multi, gating multi, MMV.
(c) *Bistable:* bistable multi, bistable (by itself), flip-flop, trigger, Eccles-Jordan, binary, toggle, scale-of-two, BMV.

The basic operating principle of the transistor multi family is illustrated by Fig. 3.1. It is the transistor equivalent of the anode-coupled multi originally developed with valves. The output from the collector of a transistor Q is fed through an impedance Z to the base input of another transistor Q'. This second transistor gives

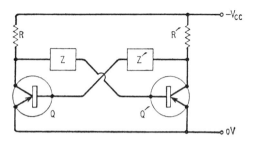

Fig. 3.1. Basic transistor multivibrator circuit

50   *Elements of Transistor Pulse Circuits*

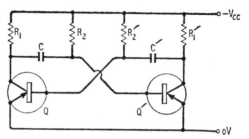

Fig. 3.2. Elementary astable multivibrator

out from its collector a corresponding amplified and phase-inverted output which is fed back through an impedance Z' to the input of the first transistor, Q. The signal is now further amplified and phase-inverted through Q, so that an initial signal at the collector of Q reappears at the same point greatly amplified and in the same phase. The circuit will therefore tend to go into violent oscillation. What actually happens depends upon the nature of the cross-coupling impedances Z and Z'. If they are both capacitances, the multi will be astable. If one is a capacitance and one a resistance, the system will be monostable. If both cross-coupling impedances are resistances, the system is bistable. This chapter deals with the first of these circuits.

THE ASTABLE MULTIVIBRATOR

The astable multi is a self driven oscillator which generates a train of rectangular waves, and, since it requires no triggering signals, it is widely used as a primary source of pulses. The basic transistor circuit is illustrated in Fig. 3.2 (note that both cross-coupling impedances are capacitances). Other astable multis using different feedback circuits are possible, but the collector-coupled version is by far the commonest and so we use it to illustrate the features of the basic circuit.

Operating principle of astable multi

To understand how the circuit of Fig. 3.2 works, the waveforms given in Fig. 3.3 are needed (with negative voltages upwards for p-n-p transistors). The sequence of operations is as follows. When the circuit is first switched on, both transistors tend to conduct because current is supplied to the bases through R_2 and R'_2. The

Astable Multivibrators 51

positive feedback from output to input takes the circuit into oscillation between states where one transistor is on and the other off. At the end of period (1) in Fig. 3.3, the base voltage of the left-hand transistor Q is rising towards zero volts. As the base passes through zero and rises negatively, Q starts to conduct and causes the collector voltage of the transistor to drop towards zero. This voltage drop is transmitted to the base of the other transistor Q' via the coupling capacitor C and causes a rise in Q' collector voltage. This rise is applied back to the base of Q via the coupling capacitor C', so causing a further increase in the collector current in Q. The action develops cumulatively and causes the base voltage of Q to rise rapidly to a small negative value and its collector voltage to drop close to zero. At the same time the base of Q' drops to a large positive value (well beyond the Q' base cut-off value), and its collector rises close to the supply voltage. Note that the base of Q cannot

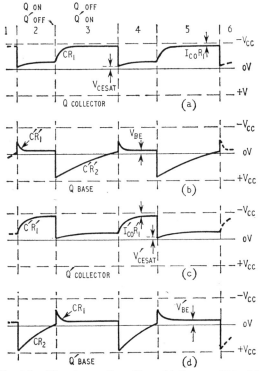

Fig. 3.3. Waveforms of astable multivibrator of Fig. 3.2

rise above the small negative value because the flow of base current in the transistor input makes the base-emitter resistance small and limits the voltage developed across it. In this rapid switch over, the circuit is said to 'flip'.

The regenerative switching stage is now complete and the circuit is then said to 'relax' during period (2) in Fig. 3.3. (This is the origin of the description 'relaxation oscillator' sometimes applied to this multi.) During this period, two things happen. First, C' *charges* via R'_1 and the low resistance base-emitter path of Q, thus causing the base of Q to return towards zero with a time constant $C'R'_1$ and the collector of Q' to rise towards the rail voltage V_{CC}, with the same time constant. Secondly, C *discharges* through R_2 and the base voltage of Q' rises towards zero with a time constant CR_2.

As soon as Q' base rises through zero at the end of period (2), Q' starts to conduct, the sudden regenerative action takes place all over again and the circuit 'flops' through the unstable transition state once more. Regeneration ceases at the beginning of stage (3) when Q is cut off by the base going positive. Stage (3) now sees a further quiet or relaxed period when C *charges* via R_1 and the low resistance base emitter path of Q', causing Q' base voltage to drop towards V_{BE} with a time constant CR_1, and Q collector voltage to rise towards V_{CC} with the same time constant. Also during stage (3), C' discharges through R'_2 and the base voltage of Q rises towards zero with a time constant $C'R'_2$. When Q reconducts, sudden amplification and regeneration take place again, the circuit flips through the unstable transition stage once more and the cycle repeats itself.

To make the rise in collector voltage to V_{CC} rapid, the time constants CR_1 and $C'R'_1$ should be short. For this, it is usual to make the collector load resistors R_1 and R'_1 fairly low values, and to keep C and C' as low as possible. The capacitor values cannot be selected without restriction, however, as they are also dependent on the repetition rate (oscillation frequency) desired.

Oscillation frequency of astable multi

The frequency at which the astable multi oscillates is set principally by the time constants CR_2 and $C'R'_2$ in Fig. 3.2, although it is also affected slightly by other factors such as the collector leakage current, I_{cbo}, and the base-emitter drive voltage, V_{BE}, of the transistors used. Also for high frequency operation, transistor internal and circuit

Astable Multivibrators

stray capacitances become significant. To a first approximation, however, at relatively low frequencies it can be shown that the period of oscillation is expressed by

$$T(\text{seconds}) = 1/f = 0\cdot7(CR_2 + C'R'_2)$$

Square wave generator

In the circuit described above, the duration of the negative and positive going parts of each collector waveform is unequal, i.e. the circuit is 'asymmetrical'. It is possible, however, to make the two

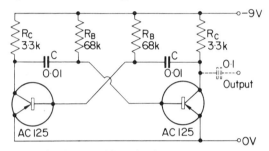

Fig. 3.4. Simple 1 kHz astable multivibrator or square-wave generator

parts of equal duration, if the time constants CR_2 and $C'R'_2$ are made equal. Most commonly the circuit is made completely symmetrical with $C = C'$, $R_1 = R'_1$ and $R_2 = R'_2$. The circuit is then known as a 'symmetrical' multi or 'square wave generator' because of the collector waveform. The period of oscillation becomes $1\cdot4\ CR$ and the frequency $1/(1\cdot4\ CR)$.

Practical design of astable multi

A precise paper design of an astable multi is quite complex, but the simple practical 1 kHz square wave generator shown in Fig. 3.4 can be used to illustrate at least the more important design points.

The voltage ratings of the transistors can be worked out from the waveforms of Fig. 3.3. They must have a V_{ce} rating at least equal to the d.c. supply voltage. Also the bass-emitter junction sees a peak reverse voltage equal to the supply voltage and the collector-base junction nearly twice this. In the circuit of Fig. 3.4 this means V_{ce}, V_{cb}, V_{eb} ratings of 9 V, 18 V, 9 V respectively. The AC 125, a conventional p-n-p germanium audio transistor, has ratings better than this.

54 *Elements of Transistor Pulse Circuits*

The battery voltage is selected at 9 V in Fig. 3.4 because this is fast becoming a practical semi-preferred value for transistor low level circuits and a wide range of standard dry batteries is available at this voltage.

The value of the collector resistor, R_C, is decided by the current to be switched. Now the current gain of most modern germanium audio transistors tends to fall off somewhat below 1 mA. We have therefore chosen a collector load resistor of 3,300 ohms, because this means that when a transistor is switched hard on and its collector is down nearly at positive rail voltage, the full rail voltage dropping across 3,300 ohms gives a collector current of just under 3 mA. (This is, incidentally, the mean current taken by the circuit from the supply, since at any point in time either one or other transistor is switched on.)

The base resistor values are then selected to ensure that the lowest gain transistor is switched firmly on. Now the AC125 has a minimum current gain of 30, so that if we take a base resistance some twenty times the collector resistance, we will ensure that the transistor bottoms firmly when switched on, because the actual base current will be more than one thirtieth of the desired collector current. The preferred value of 68,000 ohms is therefore used.

For a frequency, $f = 1,000$ Hz, the timing capacitor values, given by $1/(1 \cdot 4 f R_B)$, where R_B = base feed resistance, work out at approximately 0·01 μF.

The output from an astable multi can be taken from either collector. The load resistance is then in parallel with the collector resistance, but it has only a second order effect on the repetition frequency because the collector resistor does not appear in the approximate frequency formula. In Fig. 3.4, the output can be taken off from the right hand collector (through a 0·1 μF d.c. isolating capacitor if desired), and this simple circuit makes a useful 1 kHz square wave test oscillator, with an output impedance of 3,300 ohms.

High frequency limitations

The simple astable multi circuit shown cannot be made to operate above a certain repetition rate. Up to about 20 kHz a conventional audio transistor such as the AC125 can be used satisfactorily. Up to 100 kHz an alloy r.f. transistor can be used. Above this frequency recourse must be made to alloy diffused or other u.h.f. types, but precautions are necessary to ensure that the emitter-base voltage ratings of these are not exceeded because they are usually somewhat limited. In the audio and supersonic range the transistor is usually the limiting factor in speed of operation. It is only when we get into the r.f. range that circuit strays become significant compared with the transistor limitations.

The switching speed of the transistor itself is limited by three things—hole storage, interelectrode capacitances and fall-off of

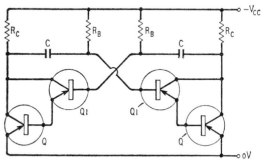

Fig. 3.5. Very low frequency astable multivibrator

current gain with frequency. Special techniques such as non-saturated switching (e.g. diode clamping of transistor collectors) can be used to increase the available repetition rate, but in general it is difficult, with any given transistor, to increase the repetition rate more than two or three times in this way.

Low frequency limitations

In the low frequency direction, the repetition rate of a simple astable multi circuit of the type shown is also limited. Since the repetition rate is proportion to $1/CR_B$ to get very low frequencies we must increase the CR_B product as far as possible. Now R_B cannot be increased above a certain value fixed by R_C and the current gain of the transistors. If we want to reduce repetition rate beyond a certain point, we can do this only by increasing the cross coupling capacitors, C. Unfortunately, to do so in practical circuits we are compelled to use electrolytic capacitors, and above a certain value of these we find the leakage currents upset the timing constants. This prevents reducing the repetition rates any further. Transistor leakage currents, too, become significant and make it difficult to get very low repetition rates. However, one effective way exists of reaching lower frequencies; this is to increase R_B

rather than C. This we can do if we can get transistors with sufficiently high current gain. Single transistors with current gains in excess of a hundred are not freely obtainable commercially, but a simple Darlington pair such as shown in Fig. 3.5 gives a compound current gain of the order of a thousand. This makes possible base resistors an order of magnitude higher than feasible with single transistors, and thus extends the lower frequency limit of the simple multi by a factor of 10 at least.

We have been illustrating the principles only of the astable multi by the simplest possible circuit, but in practice real circuits are much more complex. Frequently, to define the voltage levels accurately, and to give the fastest possible rise and fall times, the circuit is operated in a non-saturated mode. This means that the transistor is not allowed to switch hard into bottoming, with resultant long hole-storage and switch-off times. Another variant often found is to supply the direct voltage to the bases from a separately controlled rail. By varying this base supply voltage it is possible to vary the pulse repetition rate easily. Finally, we have shown only one version of the basic multi circuit, i.e. with collectors cross-coupled by capacitors. Other circuits such as the emitter-coupled equivalent of the cathode-coupled valve circuit will also be found.

Astable multis are not only used as rectangular or square wave pulse generators, but they also feature as frequency dividers in the form of the triggered astable multivibrator.

Synchronised astable multivibrator frequency divider

Free-running multis possess rather poor frequency stability since changes in supply voltage, capacitances, resistances and the characteristics of transistors affect the controlling time constant. However, the multi can be locked in step with an external signal injected into one of the transistor bases and the multi adjusted to run at a slightly slower frequency than the applied signal.

The resultant waveforms from such a procedure are shown in Fig. 3.6. Negative going sync. pulses are applied to the base of one transistor (say Q of the simple astable circuit of Fig. 3.2) at a faster repetition rate than the free-running frequency of the circuit which is indicated by the continuous line in (a) and (b). If pulse No. 1 arrives at an instant when Q is already switching on, pulse No. 2 will arrive at the transistor base when the voltage there is rising towards zero, but has not quite reached a level to switch the transistor on by itself. The negative pulse takes it immediately

Astable Multivibrators 57

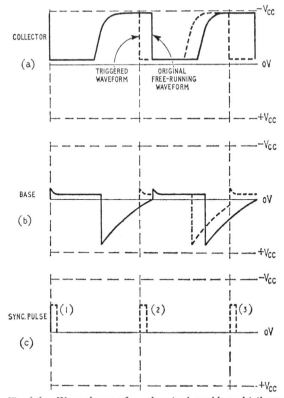

Fig. 3.6. Wave-shapes of synchronised astable multivibrator

over to a negative level and so accelerates the transition of Q into its conducting state as shown by the dotted traces, coinciding with the leading edge of pulse No. 2. In effect this causes Q to conduct before it would have done under the normal CR discharging conditions. As each successive sync. pulse causes this to happen, the astable multi locks in with the sync. pulses.

Instead of making every sync. pulse switch the free running multi, it is possible to arrange for it to be switched by every nth pulse. This means that the multi frequency locks in at $1/n$th of the input frequency. For example, an input with a repetition rate of 1 kHz can cause the multi to oscillate at 200 Hz by arranging for it to be switched by every 5th pulse. Frequency division thus takes place, and by careful design, it is possible to divide reliably up to as much as ten times. To do this, the astable multi is designed asymmetrically

58 *Elements of Transistor Pulse Circuits*

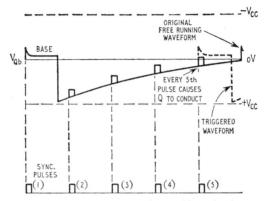

Fig. 3.7. Frequency division by astable multivibrator

so that one transistor conducts for less than the repetition time of the sync. pulses and the other for longer than the time of (*n*-1) pulses, the overall time being longer than *n* times the period of the sync. pulses. Fig. 3.7 illustrates how division then takes place. Negative sync. pulses are supplied to the base of the transistor with the short conducting time. At the fifth pulse, the transistor is prematurely switched on and thus the circuit brought into synchronism with a frequency 1/5 of the sync. pulse repetition rate. It will be noted that the amplitude of the sync. pulse is critical. If it varies significantly, the switching can be early or late and the frequency division becomes uncertain.

One convenient fact is that either a sine wave or a square wave can be used as the sync. signal and flat-topped square or rectangular output pulses obtained.

The example shown has been for an unsymmetrical output. If a symmetrical output is required at the lower frequency, it will be necessary for sync. pulses to be applied to both transistors. This can conveniently be done by including a small resistor between the commoned emitters and positive earth in Fig. 3.2 and applying the sync. pulses across this resistor.

CONCLUSIONS

As one of the basic building blocks of pulse circuit work, the astable multi deserves deeper study than is possible within the confines of this chapter. For a fuller and more rigorous treatment, readers are recommended to consult *Junction Transistors in Pulse Circuits*, P. A. Neeteson, Cleaver-Hume Press 1959.

CHAPTER 4

Monostable Multivibrators

A 'Gallup Poll' conducted by the author on a small random cross section of electronic engineers brought out interesting facts about the monostable multivibrator. Everyone interviewed had *heard* of the circuit, but only 20% could have obtained a pass mark in an examination for the question 'Describe in general terms the principle and uses of the monostable multivibrator'. Moreover, only 5% (one lone engineer in the twenty canvassed) could actually do even a simple approximate design of a general-purpose monostable. Considering the importance of the circuit for generating, reshaping, stretching or delaying pulses, for gating other circuits and for frequency dividing, it is surprising how few engineers have a working knowledge of it. This chapter aims to fill this gap with particular reference to the modern transistor versions of the monostable multi (which have largely superseded the valve versions formerly used).

In cold scientific terms, a monostable multi is a two-state device with one permanently stable and one quasi-stable state. A suitable trigger pulse will induce a rapid transition from the stable to the quasi-stable state. The multi then remains in this second state for a predetermined time which is long compared with the time of transition between states. Finally it switches rapidly back to the stable state without any further external triggering. In another way, the circuit is said to 'flip' on under a trigger pulse and 'flop' off by itself after a time T which is essentially independent of the trigger pulse characteristics and is determined only by the RC time constant of the circuit itself.

Because it requires only a single trigger pulse to go through its cycle, the monostable multi is often referred to as a 'one-shot',

60   *Elements of Transistor Pulse Circuits*

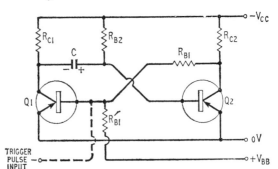

Fig. 4.1. Basic collector-coupled circuit

'single-step multi' or 'univibrator'. Because it produces a rectangular output pulse which can be used to gate other circuits, it is sometimes described as a 'gating' multi. Again, the output pulse can be differentiated to give a pulse at its trailing edge which occurs at a preset time after the input trigger pulse, and the monostable is therefore often called a 'delay' multi. The names are all suggestive of the uses to which the circuit is put.

BASIC COLLECTOR-COUPLED MONOSTABLE MULTIVIBRATOR

Transistor monostable multis fall into two classes:

(a) *Collector-coupled*, where the cross-couplings are from collector to opposite base, and
(b) *Emitter-coupled*, where one cross coupling is from collector to base and the other is by feedback through an emitter resistor common to both stages.

The collector-coupled version is analogous to the valve anode-coupled multi, and the emitter-coupled to the valve cathode-coupled one.

The basic circuit of the first of these, the collector-coupled monostable multi, is shown in Fig. 4.1. In the absence of external excitation Q2 is held on by the base current through R_{B2}. Now when Q2 is on, its collector voltage is close to 0 V, the transistor being bottomed. As a result, Q1 is held off by the network R_{B1} and R'_{B1} between 0 V and $+V_{BB}$, the potential of the base of Q1 lying somewhere between 0 V and $+V_{BB}$, depending on the ratio of R_{B_1} to R'_{B_1}. If either transistor is triggered out of its stable state, regeneration can occur due to the overcoupled positive feedback in the circuit. The resulting astable state with Q1 on and Q2 off will exist for a time determined largely by the time constant CR_{B_2}. The mechanism of this is relatively easy to follow. In the stable state with Q1 cut off, its collector is virtually up at the negative rail potential $-V_{CC}$. At the same time the base of the transistor Q2 is effectively at 0 V (if we ignore the small base-emitter volts drop in that transistor). Thus the capacitor C is connected effectively between the negative rail, $-V_{CC}$ and the earth line at 0 V, and as a result is charged up to $-V_{CC}$. When Q1 is pulsed on, its collector drops to 0 V, carrying down the left hand side of the capacitor C to that level. The right hand side of C is consequently taken immediately to $+V_{CC}$ volts. This takes the base of Q2 positive and cuts off the transistor. C then begins to discharge through R_{B_2} and its right hand side gradually rises towards 0 V with a time constant CR_{B_2}. After C has discharged sufficiently, Q2 begins to conduct again and regeneration immediately sets in to turn Q2 rapidly on and Q1 rapidly off.

The resultant approximate waveshapes are shown in Fig. 4.2. The output can be taken from the collector of either transistor, but normally it is taken from the collector of Q2. The waveshapes show

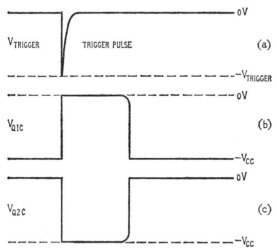

Fig. 4.2. Waveforms of circuit of Fig. 4.1. Trigger pulse (a), Q_1 collector (b) and Q_2 collector (c)

that the voltage output at the collector of Q2 is a negative going pulse.

The precise design of a collector-coupled monostable multi can be complex, but the general lines can be covered with reference to Fig. 4.1. V_{CC} is first selected equal to the output pulse voltage required. R_{C_2} is then chosen to give the output impedance required for the pulse. The transistor Q2 is selected with a collector-emitter voltage rating (at the full collector current, V_{CC}/R_{C_2}) not less than V_{CC}. When the circuit is flipped on, the base of Q_2 (connected to the right hand side of C) is carried to approximately $+V_{CC}$, and Q2 must have an emitter-base voltage rating not less than V_{CC}. Finally, since the base of Q2 switches to $+V_{CC}$ and its collector to $-V_{CC}$ when it is first switched off, Q2 must have a collector-base voltage rating not less than $2 \times V_{CC}$. R_{B_2} is chosen sufficiently low in value to ensure that Q2 is well bottomed when switched on. To make this possible, R_{B_2} should be selected to give a base current of at least twice that required to bottom Q2, i.e., $R_{B_2} = V_{CC}/2R_{C_2}h_{fe}$. For reasons which cannot be gone into here, the V_{BB} rail voltage should be selected not less than one-third of V_{CC}. R_{B_1} is normally selected to be not less than ten times R_{C_2} to ensure that it does not load R_{C_2} excessively. R'_{B_1} is then chosen to ensure that Q1 base goes more than 1 V positive in the cut-off condition. For this it should not be greater than $(V_{BB} - 1)R_{B_1}$. The transistor Q1 is then selected with a collector-emitter voltage rating better than $-V_{CC}$, because its collector travels between 0 V and $-V_{CC}$ and its emitter stays at 0 V. In this case also, because there is no capacitive cross-coupling to the collector of Q2, the emitter voltage rating is unimportant, and provided it exceeds a few volts there should be no trouble. The collector-base voltage rating need not, as a result, be more than a few volts greater than V_{CC}.

The current gain of Q1 should be selected so that the base drive provided by R_{B_1} will bottom the transistor in its on condition. R_{C_1} should not be of too great a value or it will adversely effect the recovery time of the circuit after pulsing. Very frequently R_{C_1} will be found approximately equal to R_{C_2}, the circuit being symmetrical in this respect.

The length of the on-pulse is set in the main by the time constant CR_{B_2}. An approximate formula for the pulse length time is:

$$T = 0{\cdot}7CR_{B_2} \tag{4.1}$$

For a rigorous treatment of the design of a collector-coupled monostable multi, readers are recommended to consult *Junction Transistors in Pulse Circuits*, by P. A. Neeteson, Cleaver-Hume Press Ltd., 1959.

TYPICAL MEDIUM-SPEED COLLECTOR-COUPLED MULTIVIBRATOR

In Fig. 4.3 will be found the practical circuit of a 20 μsec collector-coupled monostable multivibrator, based on the simple general design analysis given above. In the standby condition transistor Q1 is biased off and Q2 on. Q1 is triggered on by a pulse applied to its base through the 220 pF, 1 kΩ differentiating network. Diode D is

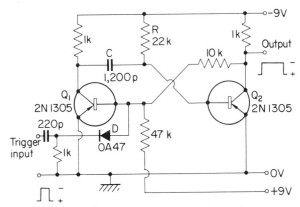

Fig. 4.3. Typical medium-speed circuit, giving 20 μsec pulse

connected so that only negative going pulses pass through to the base of Q1. The diode used, an OA47, is a low-resistance fast diode with low hole-storage properties, but any good point-contact or gold-bonded germanium diode will do. Transistors Q1 and Q2 are standard alloy germanium r.f. switching transistors, with a minimum frequency cut-off of 8 MHz. The 2N1305s were used in the circuit because they have a guaranteed minimum d.c. current gain of 50 at the 10 mA level switched in this circuit, but any similar germanium alloy switching transistor with a typical c_{hfb} of 15 MHz would be suitable.

The circuit has been shown with separate positive and negative d.c. supplies. A single-supply version can be obtained by connecting a 1,000-ohm resistor shunted by a 0·1 μF from the 0 V to +9 V rail and connecting a single 18 V supply between the +9 V and −9 V rails.

64 Elements of Transistor Pulse Circuits

TYPICAL HIGH-SPEED MONOSTABLE MULTIVIBRATOR

A transistor requires a finite time to switch on and off, and special transistors have to be used when pulses narrower than a few microseconds are required. Fig. 4.4 illustrates a typical high-speed monostable capable of producing a narrow pulse of about half a microsecond duration. For these narrow pulses, a standard germanium alloy transistor would be useless and resort must be made to diffused-base transistors.

In the state before the monostable is triggered, the right hand transistor Q2 is biased on by the current through the base resistor R

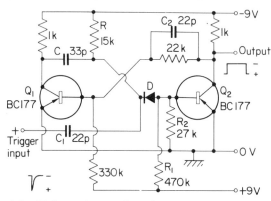

Fig. 4.4. High-speed circuit for pulses down to 0·5 μsec width

and forward biased diode D. At the same time Q1 is biased off by the 22 kΩ, 330 kΩ potentiometer across its base. On the application of a positive trigger pulse voltage to the input through C_1 across R, this voltage cuts off the diode D and causes Q2 to start switching off. Regenerative action through the cross-coupling resistor from the collector of Q2 to the base of Q1, tends to drive the latter transistor on, and the circuit flips into the quasi-stable state with Q1 on and Q2 off. During this state when the output pulse is being produced, the three capacitances C, C_1 and C_D (the capacitance of the diode D) charge up with a time constant approximately $(C + C_1 + C_D)R$, and the pulse length is $0·70(C + C_1 + C_D)R$.

The purpose of the diode D is to protect the base-emitter junction of Q2 from exceeding its voltage rating which is of the order of 1 V only (this being typical of the diffused-base type of germanium transistor). The diode D connected in series with the base of the

Monostable Multivibrators

transistor Q2 stops the positive signal up to 9 V developed at the right hand side of the capacitor C during the quasi-stable state from being applied to the base of the transistor Q2. In this connection, an additional feature of the circuit not found in simpler circuits is the potentiometer network R_1, R_2 which maintains transistor Q2 emitter-base cut-off at a low voltage during the quasi-stable period. The other additional feature is the so-called 'speed-up' or 'commutating' capacitor C_2 across the 22 kΩ cross-coupling resistor. The purpose of this, which will be examined in greater detail in a later chapter on the bistable multi, is to ensure a quasi-voltage initial drive to the base of Q1 when it is being turned on. This speeds up the transition between the stable and unstable states. The BC177 transistors used in the illustrative circuit are silicon planar diffused types with typical frequency cut-offs of 150 MHz, voltage ratings greater than 20 V and minimum current gains of the order of 50. The diode D can be any fast point-contact or gold-bonded germanium diode.

TYPICAL LOW-SPEED COLLECTOR-COUPLED MULTIVIBRATOR

Long-time-constant monostables present certain difficulties in design, particularly at high currents. This is because the timing resistor is also the resistor that provides the base drive for the normally-on transistor. For high currents, this base drive resistor must be low value. This then means that for a long output pulse the timing capacitor must be very large and in very large capacitors (which must be electrolytic) leakage currents tend to upset the timing constants. The practical circuit of Fig. 4.5 shows one method of getting round this difficulty. The particular circuit shown is designed to operate a high-current relay for the order of seconds to tens of seconds. It is basically a conventional collector-coupled monostable multi in which the timing is set by C, R_1 and R_2. When the switch S is open, the compound Darlington-pair transistor Q2-Q3 is held on by the base current through R_1 and R_2 and the relay R_{LY} is held in. When the start switch S is closed, a negative pulse is applied to the base of Q2 and initiates a flip to the state where Q1 is hard on and Q2-Q3 cut off. C then discharges through R_1 and R_2 and eventually Q2-Q3 comes on again and Q1 cuts off to return to the stable state. The circuit is thus a timer which switches off a relay for a predetermined time given approximately by

$$T = 0.7C(R_1 + R_2)$$

66 Elements of Transistor Pulse Circuits

With the component values shown, this provides a relay switch-off time which can be varied from approximately 0·5 to 5·0 seconds. The resistor R_3 is designed to reduce leakage current in Q2-Q3 in the cut-off condition. The diode D1 is a blocking diode to cut R_3 out of the timing circuit, when the diode is reverse-biased during Q2-Q3 cut off. The feed network to the start switch, S, i.e. 1 kΩ 220 kΩ, and 0·1 μF, is designed to provide an instantaneous sharp pulse for driving Q_1 on, but it has a sufficiently high d.c. resistance not to hold it on beyond the monostable pulse time if the starting switch S_1 is accidentally kept close after starting the cycle. The diode D2 and the 120-ohm resistor in series across the relay are designed to clamp the collectors of Q2 and Q3 so that, on switch-off, the inductive switch-off voltage from the relay coil cannot take the collectors substantially above the −12 V rail, thus protecting Q2-Q3 from breakdown due to excessive voltage. The 12-ohm resistor R_4 provides a stand-off voltage for the transistor bases and enables a single battery supply to be used. Because of the low value of this resistance it is not necessary normally to decouple it. Q1 and Q3 are standard miniature intermediate power transistors operated in free air without being mounted on a heat sink. Q2 is a standard high gain audio transistor used in combination with Q3 to provide a composite very high gain transistor pair which enables reasonably high values of

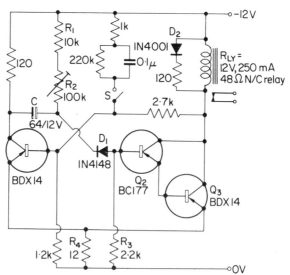

Fig. 4.5. *Low-speed multivibrator driving a relay*

Monostable Multivibrators

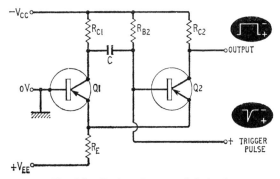

Fig. 4.6. Basic emitter-coupled circuit

resistance to be used for R_1 and R_2. By using the composite transistor, values of $(R_1 + R_2)$ at least ten times higher than with a single transistor can be used for any given pulse length and correspondingly lower timing capacitor values. Preset relay operation up to several minutes can be achieved with this arrangement.

EMITTER-COUPLED MULTIVIBRATORS

The circuits described so far have been of one main category, collector-coupled monostables. The other main category is the emitter-coupled multivibrator. Fig. 4.6 illustrates the basic circuit of this. Here Q2 in the absence of external triggering, is normally conducting due to the base current supplied by R_{B2}. R_E is chosen sufficiently high in value to ensure that the on current through Q2 and R_E biases the emitter of Q1 more negative than the 0 V, at which the base is held, and thus keeps Q1 cut off. A positive-going trigger pulse applied to the base of Q2 is followed by the emitters of both Q2 and Q1, until the emitter of Q1 goes positive with respect to its base. Q1 then begins to switch on, its collector voltage begins to drop towards 0 V and this positive voltage drop is communicated back through C to the base of Q2. Regeneration sets in and the circuit switches rapidly into the condition where Q2 is cut-off and Q1 is hard on. Q2 is held off by the charge voltage across C, applied to its base. C discharges through R_{B2} with a time constant CR_{B2}. The pulse duration before the circuit flops back to its original quiescent state is approximately $0{\cdot}7CR_{B2}$. The result of all this is a negative-going pulse output at the collector of Q2, of duration $0{\cdot}7CR_{B2}$. The detailed design of this sort of circuit need not be

68 *Elements of Transistor Pulse Circuits*

gone into in full because it is very similar to that of the collector-coupled multivibrator given earlier. For a very full analysis, interested readers should consult *Transistor Circuit Analysis* by M. V. Joyce and K. K. Clarke, Addison-Wesley Publishing Co. Massachusetts, 1961.

COMPLEMENTARY-SYMMETRY MULTIVIBRATORS

The circuits discussed so far have all used transistors of the same polarity. In the illustrations, they happen to have been p-n-p, but n-p-n could equally well have been used if available. The use of both types of transistor in combination makes possible a new family of switching circuits which have no valve equivalent. One form of the basic complementary symmetry monostable multi is shown in Fig. 4.7, where normally in the quiescent condition both transistors are switched hard on. The p-n-p transistor Q is held hard on by the base current through $R_{BO} + R_B$ (R_{BO} is only a limiting resistor to guard against accidental short circuit of the base of Q to the negative rail if R_B is reduced to zero). With transistor Q switched on, its collector voltage is close to 0 V. Under these conditions the n-p-n transistor Q' is also held on, this time by the base current through R'_B,

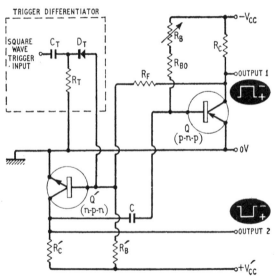

Fig. 4.7. Complementary-symmetry multivibrator, using p-n-p and n-p-n transistors (both normally on)

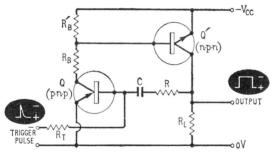

Fig. 4.8. Alternative version of Fig. 4.7 with transistors normally off

little current being deflected through R_F, since both ends of R_F are close to 0 V. When a negative trigger pulse is applied to the base of the n-p-n transistor, Q', it starts to switch off, and the voltage on its collector starts to move in the positive direction. This positive step is communicated through C to the base of Q, and tends to cut Q off, resulting in a negative step at the collector, which is communicated back through R_F and aids the negative trigger pulse applied to the base of the Q'. Regeneration sets in and both transistors rapidly switch off. The circuit then relaxes, C charging up through $(R_{BO} + R_B + R'_C)$ until eventually Q starts to switch on again and the circuit flops back into the quiescent condition with both transistors full on. The length of the pulse is set by the discharge time constant of C, and $(R_{BO} + R_B + R'_B)$, and is given approximately by $0.7C(R_{BO} + R_B + R'_B)$. $R_{BO} + R_B$ could be replaced by a single fixed resistor, if a variable pulse length is not wanted. The circuit is interesting in that it provides simultaneous outputs of opposite polarities at the transistor collectors, negative-going in Q and positive-going in Q', each with reference to 0 V.

For a 100 μsec, 6 V pulse, typical values in this circuit would be $V_{CC} = V'_{CC} = 6$ V; $R_C = R'_C = 4.7$ kΩ; $R'_B = R_{BO} + R_B = 47$ kΩ; Q = BC177; Q' = BC107; $R_F = 47$ kΩ; C = 0.0015 μF; D_T = OA47 (but any point contact or gold-bonded germanium diode is suitable); C_T 0.001 μF and $R_T = 1$ kΩ. With a pulse length of 100 μsec, the recovery time was found to be less than 20 μsec i.e., it was found possible to trigger the circuit up to a pulse repetition rate of 8 kHz.

The circuit of Fig. 4.7 has both transistors switched on during standby. For battery economy, a circuit where both transistors are cut off during standby is desirable. Fig. 4.8 gives the basic circuit of this alternative type. In the quiescent condition between pulses,

70 *Elements of Transistor Pulse Circuits*

with no current to the base of the p-n-p transistor Q, it is cut off and its collector voltage is close to the negative rail, $-V_{CC}$. With no drive voltage to its base, the n-p-n transistor Q' is also cut-off and the capacitor C is discharged. A negative trigger pulse applied to the base of Q through an isolating resistor R_T will tend to drive it full on. Its collector voltage will then fall towards the positive rail. This carries the base of Q' positive and tends to drive Q' on. The collector voltage of Q' then tends to go negative and this negative step is fed back through C and R to the base of Q, assisting the trigger pulse cumulatively until the whole circuit flips into the state where both Q and Q' are held hard on. C then begins to charge through R during the relaxation (pulse) period until eventually Q begins to switch off again and the circuit flops back into the quiescent condition, with both transistors cut off. If the on pulse is designed to be $T = CR_{eff}$, it can be shown that R_{eff} is given approximately by $R_{eff} = 0\cdot 4 \, h_{fe} h_{fe'} \times R$. The circuit has the disadvantage that the pulse time is highly dependent upon the transistor current gains, but it has the advantages that it requires no power effectively until triggered, that it can handle large load currents and can sustain long pulse periods with relatively small capacitor values. Because of the double current multiplication round the circuit loop, the timing capacitor C can be approximately h_{fe} times smaller than would be necessary in a conventional monostable multivibrator using only one polarity of transistor.

CONCLUSION

In this chapter we have taken a look at the monostable multivibrator in its transistor versions, because it is such a commonly used tool in pulse circuit work. It finds many applications in the generation of precise pulses, in reshaping pulse trains which have degenerated during transmission, in stretching narrow pulses into wider ones, in generating a time delay, in gating control circuits, and in frequency division (in the way described previously for an astable multivibrator). Most engineers find the monostable multi a satisfying circuit because so very frequently it turns out to perform almost exactly as designed theoretically and it is usually not difficult to 'get working', once you understand the basic principle on which it works.

CHAPTER 5

'Eccles-Jordan' Bistable Multivibrators

A numerologist would look at the reference 'W. H. Eccles and F. W. Jordan, *Radio Review*, 1919, Volume 1, page 143' and arrive at the 'magic number' 2 by repeatedly adding its digits together. But the reference has a more scientific connection with the numeral 2 than this. It was in this article that Eccles and Jordan first published details of their bistable multivibrator circuit, a 2-stage switching device which has since become the 'building brick' of many digital computers. Bistable multivibrators can be used to divide, count, store, or steer trains of pulses, and have become such a commonplace tool of pulse circuitry that the modern electronic engineer must have a working knowledge of the circuit and its principal properties.

The circuit has been used in many different engineering fields and has been given many different names. In this article we will call it just an 'Eccles-Jordan', because most engineers recognise it under that name. It has also, however, been described in the literature as a 'bistable', 'bistable multi', 'trigger', 'binary', 'toggle', 'scale-of-two', 'BMV' and 'flip-flip'. Sometimes again, writers use the term 'flip-flop', but purists prefer to reserve this last term for the monostable multivibrator described in a previous chapter.

Eccles and Jordan gave the original circuit in 1919 in a valve version. As it happens, the transistor version works in much the same way basically as the valve one and we will deal exclusively with transistors here.

BASIC CIRCUIT OF ECCLES-JORDAN

In simplest terms, the Eccles-Jordan is a two-stage, common-emitter, d.c.-coupled amplifier with the output d.c.-coupled back to the input. The basic circuit, stripped of inessentials, is shown in Fig. 5.1 (a) laid out as a conventional two-stage amplifier, with the forward signal going from left to right via R_{B_2} and the feedback signal from right to left via R_{B_1}. In Fig. 5.1 (b) it is redrawn in the symmetrical form normally used. This brings out clearly the hallmark of the Eccles-Jordan, the cross-coupling resistors from collectors to opposite bases. In practical circuits, these may be camouflaged by additional elements, but if you can trace an X-shaped pair of cross-coupling resistors in the circuit, it will usually be an Eccles-Jordan.

As its various alternative names imply, the Eccles-Jordan has two stable states. In each stable state, one of the two transistors is cut off and the other is held switched hard on (i.e. 'bottomed' or 'in saturation'). This built-in circuit stability which keeps either transistor permanently switched on while the other is switched off,

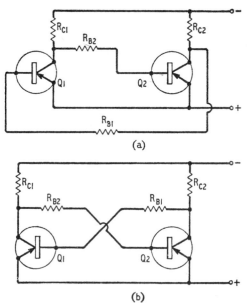

Fig. 5.1. Eccles-Jordan basic theoretical circuit; (a) drawn as conventional amplifier, (b) drawn as symmetrical circuit

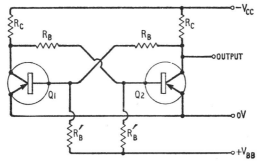

Fig. 5.2. Basic practical circuit of Eccles-Jordan with two power supplies

distinguishes the bistable from the astable and monostable multivibrators discussed in previous articles.

We now have a circuit capable of remaining indefinitely in either of two stable states. To understand the switching action, suppose that in Fig. 5.1 (b) the left hand transistor Q1 is on (with its collector voltage close to the positive rail) and the right-hand one, Q2, is off (collector voltage close to negative rail). If we apply a positive-going pulse to the base of transistor Q1, it will cut off, and the voltage on its collector will move towards the negative rail and tend to drive Q2 on *via* the cross-coupling resistor, R_{B_2}. Once transistor Q1 comes out of saturation, the circuit loop gain becomes large enough for regeneration to take control, and it switches rapidly over into the other stable condition with Q1 off and Q2 on. Because of this regeneration, during the switching interval, the external 'trigger' pulse need only initiate the switching action; the circuit will then carry the transition very rapidly through to completion on its own. We will deal more fully with triggering later below.

BASIC D.C. DESIGN

The basic circuit used to illustrate the principle of the Eccles-Jordan in Fig. 5.1 could be made to work as it stands by careful selection of transistors, but in practical applications the circuit usually appears in the form of Fig. 5.2. Here a positive rail, V_{BB}, has been added to ensure that the base of an 'off' transistor is held positive with respect to its emitter. Also the circuit has been made symmetrical with equal collector resistances, etc.

The first point to be decided in the design is the output voltage

74 *Elements of Transistor Pulse Circuits*

swing required. The output can be taken from either collector, but conventionally it is usually shown taken (as in Fig. 5.2) from the right-hand transistor, Q2. Now when Q2 is cut off, its collector is approximately at the negative rail voltage, and when it is on, its collector is at zero volts. This therefore represents the available peak-to-peak voltage output swing. The designer therefore takes a negative supply voltage. V_{CC} = required peak to peak output voltage. Commercially available germanium transistors (which are at present the most commonly used in bistable multivibrators) tend to work most satisfactorily in the lower range of d.c. supply voltages. In practice, V_{CC} will usually be found to be 6, 9, 12, or 18 volts, with a few designs going to 24 volts.

The next design point is the value of the collector current, I_C, in the 'on' transistor. In most general-purpose transistors, current gains tend to fall off below 1 mA and leakage current become significant compared with bias currents, and cannot be ignored even in an approximate analysis. Usually therefore, an 'on' collector current of not less than 1 mA is designed for. On the other hand, the higher the 'on' current, the greater the current drain from the d.c. supply. Other factors also affect the selection of the 'on' current, but in practice it will be found that I_C usually lies below 10 mA. Thus

$$I_C = 1\text{-}10 \text{ mA}$$

The selected design value of I_C automatically fixes the value of the collector resistor, R_C, because when the transistor is on, the collector resistor is effectively connected from the negative rail to zero volts. To give an 'on' current I_C, we must take

$$R_C = V_{CC}/I_C$$

The next circuit value that can be designed is the collector-base cross-coupling resistor, R_B. This must be low enough in value to ensure that the lowest-gain transistor used is fully switched on. If $h_{fe \text{ min}}$ is the lowest d.c. current gain at I_C for the transistor type used, then a base current $I_C/h_{fe \text{ min}}$ is required to saturate the transistor. R_B must be low enough to provide this base current together with the extra current diverted from the 'on' base through R'_B and also to give a margin against transistor current gain falling with time or low temperature. A common empirical rule is to take

$$R_B = \tfrac{1}{2} h_{fe \text{ min}} R_C$$

'Eccles-Jordan' Bistable Multivibrators 75

The base cut-off bias resistor, R'_B, and base supply voltage, V_{BB}, can now be worked out. The resistor value must be such that when the transistor is off, its base is not less than half a volt positive with respect to its emitter. A common practice is to design for 1 V base cut-off (although with some diffused-base transistors precautions

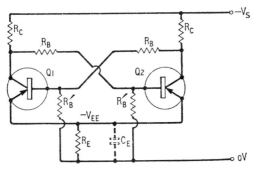

Fig. 5.3. Single-power-supply Eccles-Jordan circuit

may be necessary here). When Q1 is on, its collector is close to zero volts. The voltage at the base of the off transistor, Q2, is then set by the potentiometer R_B, R'_B across V_{BB}.

For 1 V cut-off,

$$R_B/(R_B + R'_B) = 1/V_{BB} \tag{5.1}$$

and the current through R_B is given approximately by $1/R_B$. Now, when Q2 is on, its base is close to zero volts, and current flowing down R'_B to V_{BB} diverts some of the current supplied to the base from V_{CC} via R_C, R_B. We have already selected R_B to provide double the necessary current to saturate Q2, so that up to 1/3 of the current through R_B may be safely diverted away from the 'on' base through R'_B, without bringing Q2 out of saturation. To provide a margin we assume that only about 1/5 of the available current at the base is diverted through R'_B. We therefore take $V_{BB}/R'_B = \tfrac{1}{5}(V_{CC}/R_B)$, and substituting from (5.1) arrive at the design formulae:

$$V_{BB} = V_{CC}/(V_{CC} - 5)$$

and

$$R'_B = 5R_B/(V_{CC} - 5)$$

In Fig. 5.2, separate positive and negative supplies with a common earth return have been used. In practice it is sometimes more convenient to use a single supply, and this is achieved by a circuit of the type shown in Fig. 5.3. In this, either Q1 or Q2 is switched on during the stable states and the collector current, I_C, flowing through R_E, sets up a constant negative potential at the common emitters. The d.c. design of the single-power-supply circuit can be derived from that of the double-power-supply circuit of Fig. 5.2, because they are the same circuits with $V_S = V_{CC} + V_{BB}$ and $R_E = V_{BB}/I_C$.

The capacitor C_E shown dotted across R_E in Fig. 5.3 is optional. It ensures that the d.c. conditions are not materially affected by the switching transients occurring across R_E, but often it is dispensed with. A theoretical formula for C_E can be derived showing its dependence on the transient switching times, but a convenient practical rule is to make

$$C_E = 2R_E/f_{hfb}$$

where f_{hfb} is the minimum common-base cut-off frequency in the transistor type used.

Both the circuits in Figs. 5.2 and 5.3 are saturated switching circuits; i.e. one transistor is switched hard on in the stable state. This has the drawback that a saturated transistor takes longer to switch off than one which has not been saturated. In the early days of transistor circuit design, much effort was spent in designing non-saturating circuits to get higher switching speeds with the transistors then available (which were of limited switching speed). Nowadays, on the other hand, very fast switching transistors are widely available, and for most practical circuits sufficient speed is available from saturated designs. Saturating the 'on' transistor makes the voltage swing independent of the transistor characteristics, and thus largely independent of temperature. It would seem that unless switching time is really critical, and sufficiently fast transistors are not available, one should now normally design for saturation switching.

Many diffused-base transistors have low emitter-base bias breakdown voltages of the order of 1 V. When these are used it is necessary either (a) to limit the back-bias voltage, (b) include external diodes in the emitter or base leads, or (c) design the parallel combination of R_B and R'_B of sufficiently high value to limit the current flowing in the broken down junction to a safe value.

In the above analysis, for simplicity we have ignored the effect of transistor leakage currents, particularly at elevated temperatures. These are quite considerable and can substantially modify the simplified design given. However, the procedure given is satis-

'Eccles-Jordan' Bistable Multivibrators

factory for normal room temperature operation. Readers interested in a more rigorous treatment should turn to a standard textbook such as *Design of Transistorised Circuits for Digital Computers*, by A. I. Pressman, Chapman and Hall, 1959.

SWITCHING DESIGN

So far we have concerned ourselves mainly with the d.c. design of the Eccles-Jordan. Now we must look more closely at the problem of switching between stable states. The circuit can be triggered from one state to the other by applying a suitable trigger pulse. This can be done in several ways, but they all reduce in principle to turning off an 'on' transistor or turning on an 'off' transistor, until

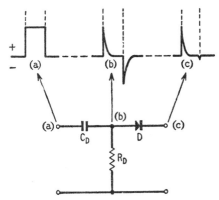

Fig. 5.4. Trigger pulse characteristics; (a) rectangular pulses, (b) bilateral differentiated pulses, (c) unilateral differentiated pulses

the circuit loop gain rises sufficiently to set up a regenerative switchover. Before we examine the various possible circuits, we might well give a little thought to the trigger pulse to be used.

In most computer switching applications, the pulses handled are rectangular waves as shown at (a) in Fig. 5.4. If such a pulse is fed into a differentiating circuit C_D, R_D, the waveshape at the output (b), becomes two sharp 'spikes', one positive and one negative corresponding to the leading and trailing edges of the initial rectangular pulse. If the differentiated pulse is then fed through a diode, D, only the positive-going spikes appear at the output, (c), the negative-going ones being almost completely blocked off. The width of the differentiated pulses is proportional to the $C_D R_D$

time constant. This circuit (when the diode is used) gives one sharp positive-going pulse for each rectangular input pulse. It is the most commonly used trigger pulse input circuit.

If we now apply the trigger circuit of Fig. 5.4 to an Eccles-Jordan as shown in heavy outline in Fig. 5.5, we can cause it to switch between bistable states. Suppose that Q1 is on, and a trigger pulse is applied to the input of C_{D_1}, it will be differentiated and appear as a sharp positive voltage spike at the base of Q1. This positive spike will tend to drive Q1 off and to initiate a switch-over to the state where Q1 is off and Q2 on. Once Q1 is off, further pulses to C_{D_1} will not cause a change of the bistable state because they will just tend to drive Q1 farther off. To cause a switch-over, we must apply a positive pulse to the base of Q2, and this we can do by a similar network C_{D_2}, R_{D_2}, D2 to the base of the second transistor. If we regard the bistable as 'set' when Q1 is off and Q2 on, and 'reset' when Q2 is off and Q1 is on, we can label the two trigger inputs 'S' and 'R' correspondingly. We can then set the bistable by a pulse to input S, and reset it by a pulse to input R. Thus we can select either of the two stable states at will. (Incidentally, on the bench a handy way of manually switching off an on transistor is merely to short its base to emitter).

We have illustrated the switch-over of an Eccles-Jordan by a positive pulse to an 'on' base. Another way is to apply a negative pulse to an 'off' base (for example by reversing D1 and D2 in Fig. 5.5), but this method is less commonly used, because a larger triggering voltage is required to overcome the cut-off potential on the 'off' base. Trigger pulses can be applied at other points in the Eccles-Jordan, such as the collector, collector-and-base simultaneously, or across common collector or emitter resistors.

STEERING CIRCUITS

In designing counters, shift registers, etc., it is often necessary to make alternate sides of an Eccles-Jordan conduct on alternate trigger pulses from a single source. There are many 'steering' circuits to achieve this. The more common of these are illustrated in Fig. 5.6. In all these trigger circuits, when a pulse arrives it finds one transistor off and one on, and one diode blocked off so that the pulse is directed or steered through the other diode to switch the Eccles-Jordan over. The next pulse finds the positions reversed and switches the circuit back again. In practical circuits, the trigger load resistors, R_D, R_{D_1} and R_{D_2} in Fig. 5.6, will often be found

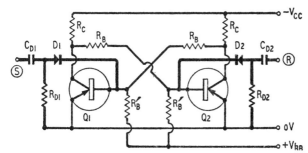

Fig. 5.5. Turn-off triggering of Eccles-Jordan

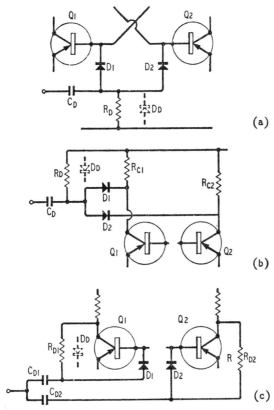

Fig. 5.6. Trigger 'steering' circuits, (a) base triggering, (b) collector triggering, (c) hybrid triggering

replaced by diodes (connected as shown dotted) to speed up the recovery time of the steering circuit between pulses. The selection of the trigger circuit capacitances and resistances, C_D and R_D, requires some thought, to ensure that sufficient drive power is obtained to switch the Eccles-Jordan, while ensuring that the recovery time of the differentiator is not so long as to limit the triggering repetition rate unduly. A detailed analysis of trigger circuits is beyond the ambit of this chapter, but some useful generalisations can be made. A fairly common rule is to make R_D approximately equal to the collector to base cross-coupling resistance.

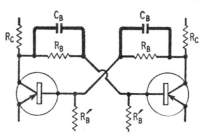

Fig. 5.7. 'Speed-up' or 'commutating' capacitors

The value of C_D should then be selected so that the $C_D R_D$ time constant is less than the repetition time of the train of pulses being steered. It should be kept as large as practicable, so that the trigger pulse does not cease before the Eccles-Jordan has begun to switch over. A useful rule is to start with $C_D R_D$ equal to one-fifth of the shortest pulse repetition time and then adjust by cut and try methods to ensure both satisfactory triggering and resolution time.

SPEED LIMITATIONS OF ECCLES-JORDAN

The switching speed of the basic Eccles-Jordan circuits so far described will be severely limited without some form of capacitor compensation. In practice, Eccles-Jordans normally incorporate small capacitors shunting the collector-base cross-coupling resistors. These 'speed up' or 'commutating' capacitors (C_B in Fig. 5.7) improve the loop gain of the circuit at high frequencies and cause faster switching between bistable states. The $C_B R_B$ combination may be considered as a compensated attenuator in conjunction with the parallel input resistance and capacitance of the transistor, $r_{b'e}$, and $C_{b'e}$, which passes a square wave without distortion. The theoretical

'Eccles-Jordan' Bistable Multivibrators

design of speed-up capacitors is complex and a complete switching time calculation is seldom worth while, because of the numerous time constants involved and because the results depend strongly on the trigger wave-shape and amplitude. Most frequently, designers start from a 'guesstimated' value of C and adjust empirically.

One starting point sometimes used is to make

$$C_B = h_{fe}/(6f_{hfb}R_B)$$

In connection with the triggering of an Eccles-Jordan, it is of interest to estimate the maximum theoretical repetition rate at which the circuit can be triggered. It has been shown (J. J. Suran and F. A. Reibert, 'Two-terminal Analysis and Synthesis of Junction Transistor Multivibrators', *Proc. I.R.E.*, March, 1956) that the maximum frequency of a multivibrator is $f_{hfb}/(h_{fe})^{\frac{1}{2}}$ permitting a maximum pulse driving rate of twice this figure. This formula neglects the effect of the transistor collector capacitance on the recovery time of any cross-coupling networks, and in practice it is found that Eccles-Jordans can be triggered reliably up to about 1/3 of the theoretical limiting frequency i.e. up to $2f_{hfb}/3(h_{fe})^{\frac{1}{2}}$. Although by no means exact, this gives a useful indication of what speed we can expect to use a particular transistor up to. For an audio transistor with a typical $f_{hfb} = 1$ MHz and a typical $h_{fe} = 100$, a maximum pulse repetition rate limit of about 66 kHz is indicated. A 10 MHz r.f. alloy transistor with $h_{fe} = 100$ could operate up to about 660 kHz without much difficulty.

One interesting point arising from the last paragraph is that the maximum switching repetition rate varies inversely as the square root of the low frequency current gain of the transistor, which suggests that low-frequency, very-high-gain transistors are the least advantageous to use in any applications.

Returning to the trigger input circuit, the lower limit of trigger power requirements can be determined by calculating the base charge in the transistor required to maintain the collector current when it is on. The trigger source must be capable of neutralising this charge in order to turn off the transistor. It can be shown that the base charge for a just saturated transistor is approximately I_C/f_{hfb}. This suggests that the least trigger power is required when we use high frequency transistors at low 'on' currents. Consequently if the trigger power is critical, it is often advantageous to use high speed transistors in slow speed circuitry.

82 *Elements of Transistor Pulse Circuits*

PRACTICAL EXAMPLES OF ECCLES-JORDAN BISTABLES

To put values to some of the design features discussed above, we now give two typical practical Eccles-Jordans with pulse repetition rates up to 10 kHz and $\frac{1}{2}$ MHz.

Fig. 5.8 gives the circuit of an economical Eccles-Jordan of the type widely used in the divider circuits of electronic organs. One interesting point is that the circuit can switch satisfactorily at the frequencies (below 10 kHz) normally handled, and yet does not use costly steering diodes. The trigger input pulse is taken *via* a $0.002\,\mu$ capacitor from one collector of the previous divider and applied across an 820 ohm common collector resistor. A signal output at half the frequency of the input pulses is taken from the left-hand transistor collector and trigger pulses at the same half-frequency are passed on from the right-hand collector to the next divider. The type AC125 used in the practical circuit are medium-gain audio transistors in TO1 encapsulations specially designed for direct printed circuit board mounting, which make possible very small boards suitable for modular assembly of an organ.

In Fig. 5.9 we have the circuit of a 'conventional' computer Eccles-Jordan capable of working up to a counting speed of $\frac{1}{2}$ MHz reliably. The trigger differentiating circuit uses a diode D1 instead of the conventional resistive load as this improves the rear

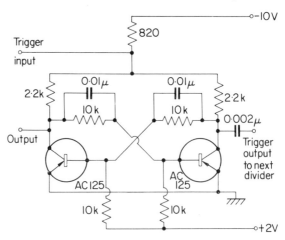

Fig. 5.8. Typical low-frequency (below 10 kHz) economical Eccles-Jordan used as divider in electronic organs

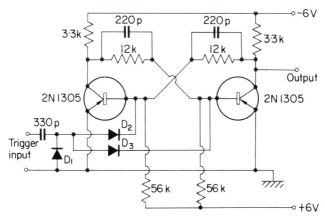

Fig. 5.9. Conventional ½ MHz Eccles-Jordan computer bistable

flank of the trigger pulses, and so makes possible a higher switching rate. The trigger steering is conventional base cut-off via the diodes D2 and D3. The 2N1305 transistors are germanium alloy devices with a minimum frequency-cut-off of 6 MHz.

CONCLUSION

In this chapter, we have attempted to give a working insight into the Eccles-Jordan bistable multivibrator. Specific applications of this type of circuit will be given in later chapters, but before we leave the subject, a word of warning is necessary. Sometimes an Eccles-Jordan will be erratic in spite of everything seemingly being correctly designed. One of the most insidious causes of erratic operation is an over-sensitive circuit. For example, it is possible for the circuit to be so sensitive that it switches not just the way it is intended but it also switches back spuriously on a single pulse. It may not lock out after switching on the front of a single pulse, but detect the tail of the pulse as an additional trigger signal. At the time of writing this article, the author came across a curious case where an Eccles-Jordan would operate happily at 30 kHz and continue to work down to a few pulses a second if the frequency was turned down steadily. But, if an attempt was made to trigger the circuit directly at slow speed without starting at the higher, it ceased to work. No explanation of this apparent hysteresis has been reached yet, but when you are working with Eccles-Jordans you must always be prepared to find 'hysterical' operation such as this which may not be easy to eliminate.

CHAPTER 6

Waveform Shaping

Electronics nowadays is much taken up with manipulating strings of pulses. This manipulation often involves shaping or reshaping pulses and the modern circuit engineer is expected to know something of at least the principles of the basic waveform shaping circuits to be described in this chapter.

The commonest waveforms met with are illustrated in Fig. 6.1. The basic one (from which all others can be derived by so called 'Fourier synthesis') is the sine wave at (a) in the figure. The 'step function' at (b) is merely an abrupt change between two d.c. levels. When engineers talk of 'pulses' loosely, they generally mean a rectangular pulse, as in (c), made up of two equal steps in opposite directions. If pulses such as these are in a repetitive series or 'train' as at (d) they are usually referred to as 'square waves'. Exponential waveforms are exponential functions of time and may be rising as in (e) or falling as in (f). A 'ramp' waveform is usually taken to mean a nearly straight rising waveshape as in (g). The 'sawtooth' of (h) is a train of rising and falling ramps. By passing these waveshapes through combinations of passive or active circuit elements, we can modify their shapes substantially. For convenience we will look at such 'shaping' circuits in three main categories: (a) linear, (b) non-linear passive, and (c) non-linear active.

LINEAR WAVEFORM SHAPING

Linear waveshaping circuits are those which preserve the waveform of a sine wave passed through them. On the other hand, any non-sinusoidal waveshape will have its form altered. The two most

High-pass RC Filter (Differentiator)

The circuit of Fig. 6.2 (a) is a high-pass RC linear filter. It is linear because a sine wave, $E = E \sin 2\pi ft$, passed through it still reappears as a sine wave at the output, but with a reduced amplitude $E/(1 + f^2_1/f^2)^{\frac{1}{2}}$ and a phase angle advance of $\theta = \tan^{-1}(f_1/f)$ where $f_1 = 1/(2\pi RC)$. The output amplitude is 0·707 of the input, i.e.

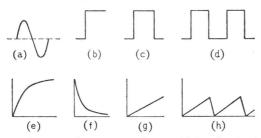

Fig. 6.1. *Waveshapes commonly met with in pulse circuits; (a) sine wave, (b) step, (c) pulse, (d) square wave, (e) exponential, rising (f) exponential, falling, (g) ramp, (h) sawtooth*

an amplitude reduction of 3 dB, at $f = f_1$. Because of this, $f_1 = 1/(2\pi RC)$ is known as the *lower 3 dB frequency* of the filter.

The response of the high-pass RC filter to an input *step voltage* is illustrated in Fig. 6.2 (b). Assume that C is initially discharged; then when the input changes abruptly by E, the voltage across the capacitor cannot change instantaneously, and the output must also change abruptly by E. Then the capacitor begins to charge up via R, and as it does so the output voltage falls exponentially until ultimately it reaches zero (because the capacitor cannot pass direct current). The output waveform can be shown to be represented mathematically by the equation $e_o = Ee^{-t/RC}$.

The high-pass RC filter changes the shape of a *rectangular pulse* as shown in Fig. 6.2 (c). The response of the filter to the front edge of the pulse is the same as in (b), but the back edge may arrive before the output has returned to zero and take the output only to the negative voltage point D. After this the output voltage rises exponentially to zero along DE with a time constant RC, until both input and output are again at zero voltage and C is completely discharged. As the capacitor cannot conduct d.c. the mean output

86 *Elements of Transistor Pulse Circuits*

level must be zero. Thus the area above and below the axis must be equal.

If RC is very long compared with the pulse width T, the output response takes the shape of Fig. 6.2 (d), where there is only a small droop on the top edge of the output pulse, and, on the back edge of the input pulse, the negative-going output step carries the output

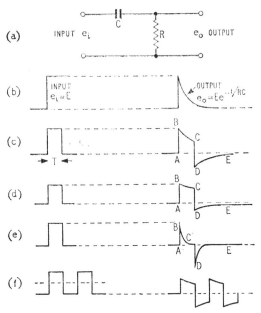

Fig. 6.2. *High-pass RC filter (differentiator);* (a) *basic circuit,* (b) *step input,* (c) *single pulse* $(RC = T)$, (d) *single pulse* $(RC \gg T)$, (e) *single pulse* $(RC \ll T)$, (f) *pulse train*

waveform only a short distance negative. The waveform then takes a long time to recover to zero because the total areas above and below the axis must again be equal.

If RC is short compared with the pulse width T, the output is as shown in Fig. 6.2 (e). Here the capacitor effectively discharges itself well before the end of the input pulse and the negative-going back edge carries the output practically to the full pulse amplitude negatively. Because of the equal positive and negative areas, the recovery curve is virtually a mirror image of the forward curve and the output is symmetrical. This process of passing a square wave through a short-time-constant high-pass RC filter is widely used for converting pulses into 'spikes' or 'pips'.

Rectangular pulses in a train of regular repetition rate or period, are often referred to as 'square waves'. Purists call them 'rectangular' waves if the *on* time of the pulse, T_1, is not equal to the *off* time, T_2, but we will follow common practice and use the term 'square wave' even where the *on* and *off* times are not equal. Engineers are interested in the steady-state response of a high-pass RC filter to such a train of square waves. If RC is very much less than both T_1 and T_2, then the output is a train of positive and negative spikes of amplitude each equal to the input pulse amplitude and coinciding with the positive and negative going edges of the input pulses. The peak-to-peak output amplitude is equal to twice the peak-to-peak input amplitude and the mean d.c. level is zero at the output. If RC is much larger than T_1 and T_2, the output pulse is virtually the same shape as the input pulse, but the capacitor gradually charges up until in the steady state the mean output level is zero. If RC is of the same order as T_1 and T_2, a complex situation exists where the output pulse is substantially a square wave but the peak droops or tilts significantly. If $T_1 = T_2 = T$ and we have a 'true' square wave, it can be shown that the output waveform becomes symmetrical about zero volts as shown in Fig. 6.2 (f) with a tilt defined as the percentage of the front edge by which the output pulse top drops) given by $P = 100\pi f_1/f$, where $f = 1/2T$ is the repetition frequency of the input train, and $f_1 = 1/2\pi RC$.

In discussing the response of a high-pass RC filter to step functions and square waves, we have cunningly assumed that the input wave is ideal, i.e. that the input has truly vertical sides—an impossibility in a real waveform. In practical circuits, with finite rise times on the input pulses, it can be shown that near the origin of time the output follows the input. Also, if the RC time constant is made smaller and smaller, the output pulse will tend to get narrower and narrower. Down to a certain value of RC the output peak will remain approximately equal to the input peak, i.e. the discontinuity in the input square wave. After that, the smaller RC the smaller the output peak. For example if RC just equals the rise time of the input wave, the peak output will be only about one-third of the peak input, but a very narrow pulse will result. Thus as we reduce RC compared with the rise time of the input pulse, the height and width of the output pulse both ultimately decrease. In practical circuits, a value of RC is chosen to give the best compromise between the conflicting requirements of sharpness and height in the output spike.

88 *Elements of Transistor Pulse Circuits*

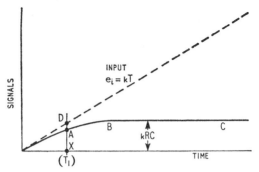

Fig. 6.3. Response of high-pass RC filter to a rising ramp input

The effect of a high-pass RC filter on a ramp input voltage is illustrated in Fig. 6.3. Here the dotted straight line represents the ramp or sweep input voltage which increases linearly with time. The output of the RC filter is as shown by the curve OABC. Near the origin, for times short compared with the RC time constant, it can be shown that the output signal falls away only slightly from the input. For example, at time T_1, small compared with RC, the departure from linearity is only AD in the diagram as compared with XD, and is commonly characterised by a so-called 'transmission error' which is the difference between the input and the output divided by the input. For times short compared with the RC time constant of the network, the transmission error can be shown to be equal to $T/2RC = \pi f_1 T$, where $f_1 = 1/(2\pi RC)$ is now the 'low-frequency 3 dB point' of the RC filter. For example, if we want to pass a 5 msec sweep through with less than 0·1% deviation from linearity, it can be shown that RC must be greater than 2·5 seconds. Now consider what happens to the output after a time T long compared with RC. This is illustrated along BC in Fig. 6.3. It will be seen that the output approaches a constant value which can be shown to be kRC, where k is the slope of the input ramp.

In general, where the RC time constant of the high-pass filter is very short compared with the time the input signal takes to make an appreciable change, the circuit is called a 'differentiator'. In this case, it can be shown that the output voltage $e_o = RCde_i/dT$, i.e. the output is the derivative of the input.

In theory the derivative of a step function is a waveform uniformly zero except at the point of discontinuity. Exact differentiation would yield an output of infinite amplitude and zero width. However, in practice the RC time constant cannot be negligibly small compared with the infinitely short rise time of a true step function, and the RC differentiator provides, in the limit of a very small RC time constant, a waveform which approaches the ideal except that the amplitude of the output peak can never exceed the input step voltage E, and the output spike has a finite width.

The differentiating action of a low-time-constant high-pass RC filter has practical uses such as measuring the rate of rise of a very fast rising pulse. Here we can assume that the leading edge of the input pulse is approximately a ramp and then the output pulse should after a very short time settle down to a constant amplitude proportional to the rate of rise of the input pulse. This output pulse peak divided by RC gives the rate of rise of the input.

Low-pass RC Filter (Integrator)

The other important linear waveshaping circuit is the *low-pass RC filter*. This takes the form shown in Fig. 6.4 (a). As its name suggests, it passes on low frequencies easily, but attenuates high frequencies, because the reactance of the shunt capacitor C falls with increasing frequency. As with the high-pass filter previously

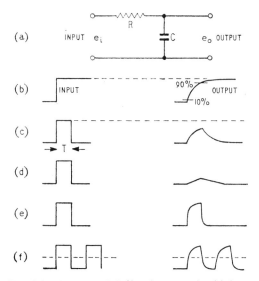

Fig. 6.4. *Low-pass RC filter (integrator)*; (a) *basic circuit*, (b) *step input*, (c) *pulse input* $(RC = T)$, (d) *pulse input* $(RC \gg T)$, (e) *pulse input* $(RC \ll T)$, (f) *pulse train*

described, this is a linear waveshaping network because a sine wave passed through it reappears as a sine wave at the output. This time the amplitude is reduced by a factor $1/(1 + (f/f_1)^2)^{\frac{1}{2}}$ and the phase angle is retarded by $\theta = \tan^{-1}(f/f_1)$, where $f_1 = 1/(2\pi RC)$. It can be hown that the 'gain' quoted for the circuit falls to 0·707 of its low requency value at the frequency f_1, which is therefore called the 'upper 3 dB frequency' for this low-pass filter. The upper 3 dB frequency of a low-pass filter, is the same as the lower 3 dB of a high-pass filter when both use the same RC product.

The response of the low-pass RC filter to an input *step voltage* is illustrated in Fig. 6.4 (b). Assume that C is initially discharged. When the input changes abruptly by E, the voltage across the capacitor cannot change instantaneously, and the output starts from zero and rises towards the steady state value E. The output waveform can be shown to be represented mathematically by the equation $e_o = E(1 - e^{-t/RC})$. It can easily be proved that the output voltage reaches 1/10 of its final value in a time equal to 0·1RC and 9/10 in a time 2·3RC. The difference between these two times (called the 'rise time', T_R, of the circuit) is the indication of how fast it can respond to a discontinuity in voltage. This rise time is related to the upper 3 dB frequency point approximately as follows:

$$T_R = 2\cdot 2RC = 0\cdot 35/f_1$$

A *rectangular pulse* is changed by transmission through the low-pass RC filter as shown in Fig. 6.4 (c). For any time less than the pulse width T, the response is the same as the response to a step input already dealt with. At the end of the input pulse, the output voltage starts to decrease to zero exponentially with a time constant RC. In the low-pass filter, since the output is directly connected to the input, the mean d.c. level of both is the same.

If RC is very long compared with the pulse width T, the output response takes the shape of Fig. 6.4 (d), where there is only a small output rise and fall as compared with the input. The output consists of exponential sections which are nearly linear. Thus with a long RC time constant we can get a good approximation to a sawtooth waveform output.

If RC is short compared with the pulse width T, the output is as shown in Fig. 6.4 (e). Here the capacitor effectively charges itself up well before the end of the pulse and on the cessation of the pulse the output returns to zero on an exponential curve.

Sometimes, when we pass a rectangular pulse through a low-pass

RC filter, instead of trying to shape it we may be trying to keep its shape as far as possible. To minimise the distortion, the time constant of the RC network should be made small compared with the pulse width. A common working rule is that the pulse shape will be substantially preserved if the 3 dB frequency is not less than the reciprocal of the pulse width. This can be put another way by saying that the RC time constant should be selected not greater than 1/6 of the pulse width. Under these circumstances, the output will rise to 90% of its final value after not more than 1/3 of the input pulse has passed, and the rectangular pulse is transmitted reasonably undistorted.

Engineers are usually concerned not so much with the response of a low-pass filter to single pulses, but to a recurrent train of them. Where RC is much less than the repetition period of the train, the output becomes a train of pulses similar in shape to the input pulses, but with rounding on the front and back edged as shown in Fig. 6.4 (f). The mean d.c. levels are the same at input and output, as are the peak-to-peak amplitudes. Where RC is much larger than the repetition period of the train, the output is a sawtooth of much smaller amplitude than the square wave input. Finally, if RC is comparable with the repetition period, the output is a very non-linear sawtooth with a marked exponential curvature on the rise and fall. Each case has one point in common—the mean d.c. level of the input and output pulse trains is the same.

Theoretically, the infinitely steep rise and fall times of the input pulses dealt with so far must be replaced in practice with something like a steep exponential ramp. The front-edge response of the low-pass filter to this waveshape is of interest because it shows what happens in actual circuits. It can be shown that if T_R is the rise time of the input waveform to 90% of its peak value, it reaches 50% of its final value in about 0·7 of T_R. If we define the 'delay' of the filter as being the time for the output to reach 50% of its peak value, it can be shown that when $RC = 0·16\ T_R$ (condition described earlier for limited distortion) the delay is about $1·7T_R$; thus the output pulse is delayed behind the input pulse by approximately T_R.

The complete response of the low-pass filter to an exponential ramp input is sometimes important. This is illustrated in Fig. 6.5 (a). If the total rise time of the exponential ramp is T and RC is small compared with this, the output (except for the delay distortion at the origin) follows the input very closely. If RC is large compared with T, the origin delay distortion is greater, and the output takes the second form shown.

The response of a low-pass RC filter to a *linear* ramp is given in Fig. 6.5 (b). If the total rise-time of the ramp is T, and RC is small compared with T, the output again follows the input closely except for the short period at the origin. The transmission error, defined as before as the difference between the input and the output divided by the input, works out at $D = RC/T$ and is relatively small. As an example, if we want to pass a 5 msec sweep with less than 0·1% error it can be shown that RC must not be more than 5 μsec. If RC is large compared with the sweep duration T, the output is very distorted as compared with the input as shown in Fig. 6.5 (b). It can be shown that in this case the output voltage e_o is approximately equal to $kT^2/(2RC)$, where k is the slope of the input ramp.

In general, where the time constant RC in the low-pass filter is large in comparison with the time required for the input signal to make an appreciable change, the circuit is known as an 'integrator'. Under these conditions it can be shown that the output is given by $e_o = (1/RC) \int e_i dt$. Where RC is large compared with T, the output gives a very good approximation to an integral. But as RC

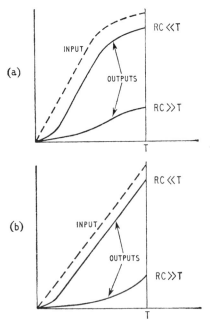

Fig. 6.5. *Low-pass RC filter response to rising ramp;* (a) *exponential,* (b) *linear*

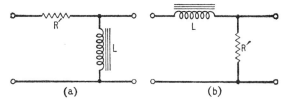

Fig. 6.6. Linear waveshaping RL filters; (a) high-pass, (b) low-pass

is reduced, the departure from true integration increases, as was shown in Fig. 6.5 (a) where the output is not in fact an integral of the input. This shows that this simple integrating circuit must be used with caution.

RL Linear Passive Networks

If in the high- and low-pass RC filters dealt with so far, we replace the resistance R by an inductance L and the capacitor C by a resistance R', we get the high-pass RL circuit of Fig. 6.6 (a) and the low-pass RL circuit of Fig. 6.6 (b). It can be shown that if the time constant L/R' is equal to the time constant RC, all the preceding analysis remains unchanged, and the properties of these low- and high-pass RL filters can be worked out easily. In actual pulse work, RL circuits are much less common than the RC ones. Where a large time constant is called for in particular, an inductance is seldom used because a large value of inductance can be obtained only with an iron-cored inductor which is physically large and expensive. In general, however, because suitable capacitor values are so much more easy to obtain than inductors, the bulk of linear waveshaping is done with RC circuits, except where special features call for RL ones.

NON-LINEAR PASSIVE WAVEFORM SHAPING

Many useful waveshaping operations can be performed if non-linear elements are added to the linear circuits already described. Space being limited, we will confine our examination of these to semi-conductor ('crystal') diodes and transistors.

Diodes as non-linear passive waveshaping elements

The ordinary diode is probably the most important non-linear waveshaping element in common use. The semiconductor diode shown in symbolic form in Fig. 6.7 (a) is a device which in one

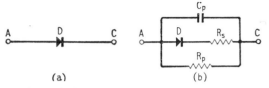

Fig. 6.7. Ideal and real diodes; (a) ideal diode, (b) real diode equivalent circuit

direction behaves as a short circuit and in the other as an open circuit. When the 'anode' A is positive with respect to the 'cathode' C, the diode conducts as though it were a closed switch. When the anode is negative with respect to the cathode, the conduction stops and the diode becomes essentially an open switch. A real diode departs from an ideal one in several respects. Firstly, the forward resistance is not zero (with semiconductor diodes it lies in the range of 1 to 500 ohms). Also this forward resistance is not constant, but varies with the applied voltage. Again, the reverse resistance is not infinite. For germanium diodes (which are most commonly used), reverse resistances of the order of 100K to tens of megohms obtain, provided the operating temperature is not above about 60°C. At higher temperatures, the reverse resistance tends to fall rapidly. There is in practice in a real diode a shunt capacitance in parallel with it which will have some effect on its circuit use. Point contact diodes have usually a capacitance of about 1 pF, gold bonded diodes about 3 pF, and junction diodes from 10-40 pF. When a diode is switched rapidly from conduction to non-conduction or *vice versa* it is found that there is a certain delay in reaching the final resistance value. This delay, called 'recovery time', is due to minority carrier storage effects in the device. In practice it ranges from nanoseconds to tens of microseconds depending on the type of diode, and becomes really important only in circuits handling fast pulses. The reverse recovery time, i.e. the time to switch from forward conduction to reverse cut-off, is usually more critical because the reverse resistance builds up gradually. It is in fact more accurate to regard real diodes as variable resistors rather than switches. Fig. 6.7(b) gives the equivalent circuit of a real diode, where R_p is the shunt reverse resistance across the diode, D is the ideal diode and R_s is the series resistance representing its forward resistance. Finally, C_p is the shunt capacitance. For many switching operations, the departure of the diode from the ideal characteristics are not very important, and in the rest of this article we will treat the diode as an ideal one with zero forward and infinite reverse resistance.

Diode clippers (or limiters)

Clipper circuits are used when we want to select for transmission that part of an arbitrary waveform which lies above or below some particular reference voltage level. Such clipper circuits are sometimes referred to as voltage selectors or amplitude selectors.

Series diode clippers operate directly in the path of the pulse or wave that is to be shaped, as shown in Fig. 6.8 (a). When the incoming signal makes the anode positive with respect to the cathode of the diode, the diode conducts, and an output voltage is developed across R_L. However, when the input voltage causes the anode to go negative with respect to the cathode, the diode cuts off and becomes an open circuit. No current then flows through the load resistor and no output voltage is developed. The result of this is shown in the waveform of Fig. 6.8 (a) where the part of the input waveform below $+V$ volts is cut off. Similar clipping can be obtained when a diode connected to a d.c. supply source is placed in parallel with the load resistance R_L across the signal input as shown in Fig. 6.8 (b). This is a parallel or shunt diode clipper. Here, when the input waveform falls below $+V$ volts, the diode conducts and the output waveform cannot fall below that level and is therefore clipped off.

In either type of clipper of Fig. 6.8 (a) and (b), the battery voltage and the diode polarity could be reversed, and then the output waveform would be clipped at the maximum instead of the minimum.

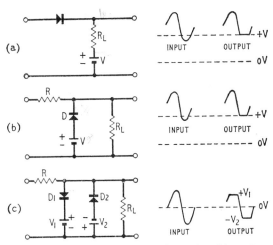

Fig. 6.8. Diode clippers (or limiters); (a) series, (b) parallel (shunt), (c) double parallel clipper (slicer)

96 *Elements of Transistor Pulse Circuits*

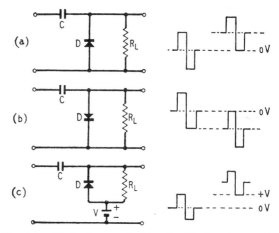

Fig. 6.9. Diode clampers (unbiased); (a) clamping negative peak to 0V, (b) clamping positive peak to 0V, (c) clamping negative to $+V$

A combination of two diodes connecting the signal line to a positive and negative d.c. supply as shown in Fig. 6.8 (c) is known as a 'double clipper' or 'slicer' and leads to both the upper and lower peaks of the waveform being cut off.

We have used the term 'clipper' in the description above but the term 'limiter' is frequently used as synonymous with it. Some writers distinguish between clipping (as referring to a voltage waveform) and limiting (to a current waveform). There is no standard practice in this matter. Where the circuit is designed to tie the upper or lower peaks of a waveform to a fixed reference voltage the clipping process is usually known as clamping. Limiters or clippers can be regarded as special clamping devices which also modify the waveshape they clamp. True clamping preserves the signal waveshape but shifts the entire waveform up or down so either the top or bottom peaks are brought to a some predetermined d.c. level.

Unbiased clamper

The capacitor-coupled clipper circuit shown in Fig. 6.9 (a) causes an input waveform to be shifted so that its negative peaks rest on zero level at the output. This does not happen on single pulses, but only on a train of pulses where after a settling down time the capacitor C charges up and shifts the mean output d.c. level.

If it is wished to clamp positive peaks of the output waveform to zero, the circuit of Fig. 6.9 (b) is used. Here again the output d.c. mean level shifts until it settles down on a train of pulses with C charged up to a level where the positive peaks of the output signal are clamped to zero volts.

Finally, if it is desired to clamp the output waveform to some voltage other than zero, the circuit of Fig. 6.9 (c) can be used. Here C steadily charges up under a train of input pulses until the output waveform has its negative peaks clamped to $+V$ volts.

NON-LINEAR ACTIVE WAVEFORM SHAPING

The non-linear waveshaping by means of diodes shown so far is very effective, but has the disadvantage that the networks have no gain and when the tops and bottoms of waveforms are removed by clipping, etc., the resulting output has a lower amplitude than the input signal. The obvious answer is to make up this attenuation by introducing amplification and this can be done by linear amplifiers after the clipper circuits. However, the same result can be achieved economically by using a transistor as a combined clipper and amplifier. When used this way the transistor stage is called an 'over-driven amplifier'. In this application the transistor is connected as shown in Fig. 6.10 (a). If a voltage waveshape is applied to the input which is of sufficiently large amplitude, the transistor will be cut off on the positive-going half cycle, and the output at the collector will rise virtually to the negative rail voltage along the path AB in Fig. 6.10 (b). Once the transistor is cut off, the input wave

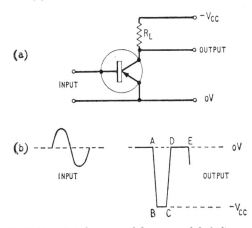

Fig. 6.10. Overdriven amplifier as amplified clipper

98 *Elements of Transistor Pulse Circuits*

carrying the base further positive has no effect on the output and the transistor collector remains clamped to the negative rail voltage as shown by BC. When the input signal moves negative, the transistor is driven hard on and the output at the collector falls virtually to zero volts along DE. As a result, a signal passed through the overdriven amplifier is clamped in one direction at the negative rail voltage and in the other at zero voltage. If it is desired to clamp the output signal at a voltage level other than zero on one side, a bias voltage can be added in the emitter circuit and the positive-going swing of the output signal cannot then pass beyond this.

PRACTICAL USE OF WAVE SHAPING TECHNIQUES

As an example of the use of some of the circuits described in this chapter, Fig. 6.11 sets out an approximate diagrammatic representation of the processes whereby in a digital computer a stable sinusoidal signal from a master clock oscillator is shaped to produce various other waveforms accurately synchronised with the input sine wave. To begin with, a train of sine waves is generated by a conventional oscillator as at (1) in Figs. (a) and (b). This sine wave output is then applied to a double-diode clipper which produced the trapezoidal waveform at (2). The sides of the waveforms are rather too sloping for accurate synchronisation so the pulse train is fed into an overdriven amplifier C which sharpens up the rise and fall times of the pulses and produces a virtually square wave output as at (3). This output may then be applied to an integrating circuit D with a long RC time constant which gives an output (4) in the shape of a sawtooth which can be used for gating purposes. The output of C can also be applied to a differentiator E with a short RC time constant to produce the output (5) in the form of a series of sharp positive- and negative-going spikes coinciding with the points at which the sine wave input waveform crosses the zero voltage axis. If this output is then applied to a negative diode clipper F it produces the waveform (6) where the negative-going spikes have been virtually removed. As a result, the waveform ends up as a train of sharp positive spikes closely synchronised with the input sine wave and available therefore as clock pulses to time the various processes in the computer.

CONCLUSION

We have not had space to cover the complete family of waveshaping circuits. The most important omissions are regenerative amplifier

Waveform Shaping 99

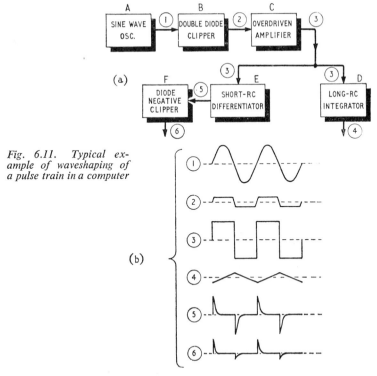

Fig. 6.11. Typical example of waveshaping of a pulse train in a computer

circuits using either transformer coupling from output back to input, i.e. blocking oscillators, or two-stage amplifiers with positive feedback from output to input, i.e. multivibrators and Schmitt triggers. We have already covered multivibrators in earlier chapters. Blocking oscillators and Schmitt triggers will be dealt with in a later chapter. Another widely used shaping circuit is the 'diode pump' and its related 'transistor pump' which are used to produce continuous or 'staircase' ramp waveforms from a train of input pulses. These too will be covered in a later chapter.

The linear and non-linear waveshaping circuits described above are worthy of more detailed exposition than is possible within the scope of this chapter. Readers interested in following them up in more detail should consult one of the two 'classics' on pulse circuits ... *Pulse and Digital Circuits* by J. Millman and H. Taub, McGraw-Hill, 1956, and *Waveforms* by B. Chance et al. M.I.T. Radiation Lab. Series. Vol. 19, McGraw-Hill, 1949.

CHAPTER 7

'Pumps' and 'Schmitts'

Two of the lesser-known but widely-used pulse circuits are the diode pump and the Schmitt trigger. The diode (or transistor) pump basically produces a 'staircase' output from a pulse train input. The Schmitt trigger is a bistable whose state depends on the d.c. level at the input. Although both are common pulse-circuit elements, it is difficult to find easily-accessible simple descriptions of their design such as are given in this chapter.

DIODE PUMPS

The diode pump basic circuit is shown in Fig. 7.1. There are two arrangements, (a) and (b), with different diode phasings. The circuit operation can be simply described as follows. Let us suppose that a string of positive pulses of amplitude V_{in} is applied to the circuit in Fig. 7.1 (a). The first input pulse will cause the capacitor C_2 to charge up through C_1 and diode D2 with a time constant equal to the product of $(C_1 + C_2)$ times the sum of the generator and diode resistances. If this time constant is small compared with the pulse duration, then C_1 and C_2 will charge up fully to V_{in} across the two before the pulse ends. By capacitor divider action, this gives an output $C_1 V_{in}/(C_1 + C_2)$ across C_2 at the end of the pulse. C_1 is usually small compared with C_2, so that the voltage change across C_2 on a pulse is small compared with V_{in}. During the pulse, D1 has been cut-off, but at the end of it, the input falls to zero, and D1 now conducts to discharge C_1. The output is unaffected during this discharge because the polarities are such that D2 is cut off. The next input pulse restores V_{in} across C_1 and C_2 in series, and repeats the small incremental charging up of C_2 as before. We thus get a series of steps to give an output 'staircase' voltage waveform.

The same sort of analysis can be applied to negative input pulses, and also, as shown in Fig. 7.1 (b) to diodes phased the other way round. The outcome of a positive pulse train will always be a

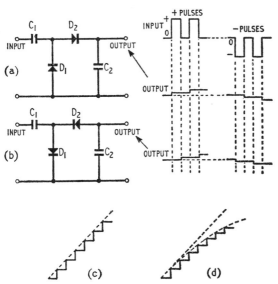

Fig. 7.1. *Diode-pump circuit and associated waveforms;* (a) *positively phased diodes,* (b) *negatively phased diodes,* (c) *linear staircase output,* (d) *non-linear staircase output*

positive-going staircase and *vice-versa*. Depending on how the diodes are phased, the output steps may coincide with the beginning or end of the input pulses. Notice that the diodes always point in the same direction . . . either both towards the output or both away from it.

Ideally, we would like to have the staircase voltage change by equal steps on each input pulse as shown in Fig. 7.1 (c). However, in the simple diode pump shown, each successive input pulse is applying V_{in} across C_1 and C_2 with the voltage across C_2 rising on each step. As a result, each successive pulse causes a progressively smaller voltage rise across C_2 and we get a drooping non-linear staircase as in Fig. 7.1 (d). It can be shown that for a train of pulses of amplitude V_{in} applied to the basic diode pump with C_2 initially discharged, the nth ouput voltage step is given by

$$dV_n = V_{in}/(1-x)x^n$$

where $x = C_2/(C_1 + C_2)$. As an example, if we take $C_1 = C_2/9$, then $x = 9/10$, and, expressed as a percentage of the input pulse amplitude V_{in}, the staircase steps rise 9%, 8·1%, 7·3%, 6·6%, 5·9%, 5·3%, 4·8% ... By the seventh pulse, the step rise has almost halved. From this analysis, it is clear that to keep the staircase reasonably linear, C_1 should be as small as possible compared with C_2. Unfortunately, the smaller we make C_1, the smaller are the output voltage steps.

For satisfactory staircase generation, fairly stringent requirements are placed on the input pulses. They should be constant in amplitude, as otherwise the output steps (which are proportional to the input pulse amplitude) will be of irregular height. They should be of large amplitude so that the output steps are not inconveniently small. They should be fairly regular, so that discharge of capacitors between pulses due to diode leakage does not introduce irregularities in the output. They should be supplied from a low impedance source and should be reasonably rectangular. One of the simplest methods of achieving all this is to feed the diode pump from a multivibrator through a buffer emitter follower.

The diodes have been assumed perfect in the above analysis, i.e. with zero forward voltage drop, infinite reverse resistance and zero switching times. In practice, for not too high-speed operation, conventional good-quality point-contact, gold-bonded or small junction diodes are satisfactory. In the case of precision requirements, high-speed, low-leakage silicon diodes may be necessary.

The capacitors should be low-leakage types, e.g. paper or mica, and their values depend on the pulse repetition rate. Usually the upper capacitor C_1 is chosen not greater than 1/20th of C_2, and the value of C_2 can be selected initially approximately as $T/20R$ where T = pulse length and R = pulse generator source resistance.

In the discussion above, no load resistance is shown on the pump output, but in practice there must be some impedance across the output, even if it is only the megohms and picofarads of an oscilloscope input. It is desirable to keep the input impedance of the stage following a diode pump as high as possible, and an emitter-follower buffer is often used for this purpose.

Apart from the various restrictions noted on the character of the input pulses, the limitations of the output load, etc., the diode pump has one basic drawback. The staircase envelope is non-linear, and the steps at the top of the 'flight' are so shallow that it may become difficult to distinguish between them.

Transistor pump

A variant of the diode pump that does give an intrinsically linear staircase is the 'transistor pump' shown in Fig. 7.2. In this, the diode across the input is replaced with the base-emitter diode of a transistor and the pump output voltage is fed back direct to the

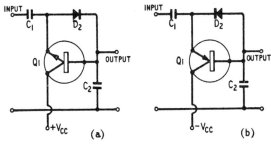

Fig. 7.2. Transistor pump circuit; (a) positively phased diode, (b) negatively phased diode

transistor base. The operation of the circuit is basically similar to that of the diode circuit, except that, when the transistor Q1 conducts, it clamps the right hand side of capacitor C_1 to the output voltage at the top end of C_2 and not to earth. This means that each input pulse is added on top of the existing output voltage and the voltage steps there will be all equal (at $V_{in}C_1/(C_1 + C_2)$). We thus get a linear staircase. This is just one further example of using bootstrapping to linearise a waveform. Two versions of the transistor pump are shown: Fig. 7.2 (a) with a positively-phased diode (and n-p-n transistor) and Fig. 7.2 (b) with negatively-phased diode (and p-n-p transistor). In each case the supply voltage, V_{CC}, of the transistor must be materially greater than V_{in}, so that it can always operate as an amplifier.

Miller integrator pump

It is also possible, as shown in Fig. 7.3, to linearise the pump staircase output by using a Miller integrator circuit. In this, the output of the simple diode pump of Fig. 7.1 is fed to a high-gain phase-inverting amplifier, with the bottom end of the output capacitor, C_2, connected to the inverter output, instead of to earth. The voltage at point 'A' remains effectively zero because any change there is phase inverted in the amplifier and applied in opposite sign at point 'B', the other end of C_2. When a pulse of amplitude V_{in} is applied to the pump

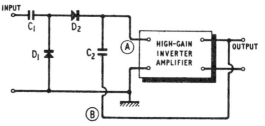

Fig. 7.3. Miller integrator linearised diode pump

input, it can be shown that C_2 will receive a voltage increment, $C_1 V_{in}/C_2$, which is independent of its previous state of charge. This means a linear staircase, of course, at the output.

Applications

The commonest application of the diode pump is to count. If to the output of the basic circuit we add a voltage comparator circuit which discharges the output capacitor C_2 whenever the staircase voltage reaches a certain level, we have a 'storage' or 'scoop' counter. (It is under one of these two titles that the diode pump is often indexed in textbooks.) The levels can be set so that after, say, five input pulses, the comparator 'fires' and thus resets the pump to zero. We then have a simple 'divide by five' circuit with the output pulse for every five input ones. The storage counter becomes a little uncertain if it is asked to divide by more than about ten at the most, but provided the pulses are fairly regular and of constant amplitude, this storage counter is an economical substitute for a bistable multivibrator counter.

The diode pump can also be used effectively as a simple staircase voltage generator, for such applications as switching the transistor base current in a series of steps in a curve tracer to display a family of characteristic curves on an oscilloscope.

A frequency meter can also be made up with a diode pump. If we shunt the output capacitor, C_2, in the pump with a resistor, R, it can be shown that for a train of pulses of repetition frequency, f, and amplitude, V_{in}, the average output voltage will settle down to fRC_1V_{in}, provided R is much less than $1/(fC_1)$ and C_2 is much greater than C_1. Now, if R, C_1 and V_{in} are kept constant, the average output voltage is proportional to the frequency. A high-impedance voltmeter (e.g. valve voltmeter) across the output can thus be calibrated directly in repetition frequency. This type of frequency-voltage converter has been widely used in frequency modulation radar systems and nuclear radiation measurements.

The diode pump can also be set up as a capacitance meter. If the above frequency meter is operated at a fixed frequency, the average output voltage, $fRV_{in}.C_1$ will be proportional to C_1. This forms the basis of a direct-reading capacitance meter, where the capacitance to be measured is inserted as C_1.

SCHMITT TRIGGER

The other type of widely used basic pulse circuit to be described is the 'Schmitt Trigger'. Under the innocuous title *A Thermionic Trigger*, Otto H. Schmitt gave, in the Journal of Scientific Instruments, 1938, Vol. XV, p. 24, the first account of an interesting bistable valve circuit he had developed. Basically it was a cathode-coupled bistable multivibrator, whose state depended only on the d.c. level at the input terminal. This proved a most useful circuit and came into common use with valves before it was adapted to transistors during the 1950's. In the transistor version, the Schmitt trigger is often given the name 'emitter-coupled binary' in textbooks, but most working engineers call it just a 'Schmitt'.

The transistor Schmitt circuit arrangement is shown in Fig. 7.4 (with n-p-n transistors so that positive voltages read conveniently upwards in the diagram). The resistor values are so chosen that normally when no voltage is applied at the input, transistor Q1 is cut off, and Q2 is conducting. If a voltage more positive than V_U (the 'upper trip voltage') is applied to the input, Q1 is driven on and Q2 switches rapidly off. So long as the input is held above V_U, Q1 remains on and Q2 off. If now the input voltage is allowed to fall below V_L (the 'lower trip voltage'), Q1 switches off again and Q2 comes on. Thus Q2 is on or off depending on whether the input voltage is low or high. The circuit exhibits 'backlash' or hysteresis in that the upper trip point V_U is above the lower trip point V_L and the circuit switches on at a higher input voltage than it switches off. The circuit is actually a regenerative bistable whose state depends on the amplitude of the voltage at the input. It belongs to the bistable multivibrator family, but the familiar 'X' of cross-coupling resistors does not appear, because, although one collector is cross-coupled to the opposite base through R_5, the other cross-coupling is by means of the common-emitter resistor R_2. Hence arises the alternative name 'emitter-coupled bistable'.

A full design taking into account backlash control, high temperature operation and switching speed is too complex for the scope of this chapter. However, a very approximate 'first order' design can be given which shows how the circuit operates and can be used to

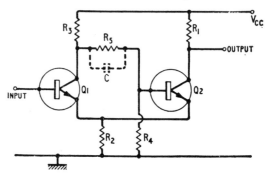

Fig. 7.4. *Schmitt trigger basic circuit*

produce practical approximate results for not too high speeds or temperatures, or too small backlash limits. Usually the designer wants to switch a peak-to-peak voltage, V_o, in an output load resistance (R_1 in Fig. 7.4) with specified input trip voltages, V_U (upper) and V_L (lower), giving an input backlash, V_U-V_L. When Q2 is on, it must pass a current $I_o = V_o/R_1$. This passes through the common-emitter resistor R_2 and gives a voltage $V_o R_2/R_1$ at the common emitters. Q1 is cut off and contributes no current to R_2. If the input voltage on the base of Q1 is taken steadily positive from zero, the circuit will begin to switch over when Q1 base voltage passes V_U. Neglecting base-emitter voltage drops (which are relatively small), this gives $V_U = V_o R_2/R_1$, and thus fixes the common emitter resistor value $R_2 = R_1 V_U/V_o$. We can now select the rail voltage, V_{cc}, to be the next standard battery voltage greater than $V_U + V_o$. To make the circuit insensitive to spreads of current gain in Q2, we make the bleeder current down R_3, R_5, R_4 ten times the maximum base current to be met with in Q2. This maximum base current will be $V_o/(R_1 h_{femin})$, where h_{femin} is the minimum d.c. current gain anticipated in Q2 at current V_o/R_1. Since Q2 is on, the voltage at its base will be approximately equal to its emitter voltage (which we have already shown to be V_U). Thus $V_U = 10R_4V_0/(R_1h_{femin})$, from which

$$R_4 = h_{femin}R_1V_U/(10V_0)$$

If we now assume that the voltage across the input has been increased just above V_U, the circuit will have switched to the state where Q1 is on and Q2 off. The current through Q1 is approximately V_U/R_2. The collector load resistor, R_3, of Q1 could now be computed by setting out the various Kirchhoff equations for the circuit, but as a simplifying assumption (which agrees with many practical circuits) we assume instead that $R_3 = R_1$, i.e. make the two collector resistors equal. If we now reduce the input voltage below V_U the circuit will eventually be on the verge of switching over to Q1 off and Q2 on, when Q1 base input voltage has fallen to V_L. The emitter of Q1 will be approximately at the same voltage, V_L. As Q2 is beginning to switch on, its base voltage also is approximately equal to V_L. This means that the current down R_4 is approximately V_L/R_4, and this is supplied down the chain R_3, R_5. Through R_3 also flows the collector current of Q1, which we saw is approximately V_L/R_2. Thus the voltage at the collector of Q1 must be $V_{CC} - R_3(V_L/R_4 + V_L/R_2)$. The voltage at Q2 base being V_L, we find by proportional division across R_4, R_5 that R_5 must be given by

$$R_5 = (V_{CC}/V_L - 1 - R_3/R_4 - R_3/R_2)R_4$$

We have thus arrived at first order formulae for all the resistance values in the basic Schmitt trigger of Fig. 7.4.

To put illustrative values to this design, assume that $V_0 = 6$ V, $I_0 = 6$ mA, $V_U = 4$ V, $V_L = 2$ V, $h_{femin} = 50$. This leads to $R_1 = 1$ kΩ, $R_2 = 680$ ohms, $R_3 = 1$ kΩ, $V_{CC} = 12$ V (allowing 2 V above $V_o + V_U$), $R_4 = 3 \cdot 3$ kΩ, $R_5 = 12$ kΩ, and Q1 = Q2 = 2N1305 (with minimum h_{fe} at 6 mA of 50).

It must be emphasised that the design outline method given is largely empirical. A full theoretical design is, of course, possible, but very complex, and usually a little unprofitable, because transistor parameters have such wide spreads in practice. The simplified design given can be used to get first approximate results, and final adjustments to meet the exact design requirements can be made by cut-and-try methods. Those interested in a rigorous detailed design should consult page 6–55 of *Selected Semiconductor Circuits Hand-*

book, by S. Schwarz, Wiley and Sons, 1960, or page 334 of *Transistor Logic Circuits*, by R. B. Hurley, Wiley and Sons, 1961.

So far we have not mentioned the capacitor C, shown dotted in Fig. 7.4. This is the 'commutating' capacitor used to speed up the circuit switchover, where the pulse repetition rate is high and fast switch over is essential. This capacitor should, with its shunt resistance, provide a time constant shorter than that of the transistor input impedance. Values in the range of 100-500 pF are common. A theoretical design value can be worked out, but common practice is to select a suitable value by trial, making it as large as possible consistent with reliable triggering at the maximum pulse repetition rate aimed at. For slow repetition rates, up to several kHz, the speed-up capacitor can usually be omitted.

The Schmitt trigger is a most attractive circuit to the designer in that its trigger sensitivity and stability can be very high. The input impedance tends to be high because of the undecoupled emitter resistor. The skeleton design given earlier did not take into account switching rates, but reliable switching repetition is easily achieved up to 100 kHz with a.f. alloy transistors, to 1 MHz with r.f. alloy types and to 10 MHz with diffused-alloy types.

Refinements

A number of refinements can be added to the basic circuit of Fig. 7.4. Firstly the amount of backlash in the circuit can be controlled by a resistor R in series with the emitter of Q2 as shown in Fig. 7.5 (a). By making R variable, the backlash can be made adjustable. R should be selected initially at about one-tenth of the common emitter resistor value and then adjusted by trial and error. Practical circuits also often include an input series resistor (R' in Fig. 7.5 (a)) to prevent over-driving of Q1. A suitable rule of thumb is to start with this equal to the collector resistor and adjust by trial and error.

Another variant of the basic circuit often met with is shown in Fig. 7.5 (b). Here the base of Q1 is connected to the slider of a variable resistor RV connected with a limiting resistor R across the d.c. supply rail. This arrangement enables the quiescent voltage on the base of Q1 to be preset so that the supplied input voltage necessary to trigger the Schmitt can be set at any convenient d.c. level. RV can conveniently be selected at about ten to one hundred times the common emitter resistor value R_E, and the limiting resistor R can be chosen as about one-tenth of RV.

Up till now we have dealt with d.c. operation of the Schmitt but

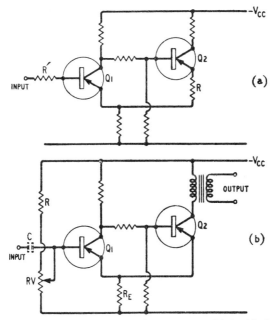

Fig. 7.5. Variants of basic Schmitt trigger; (a) backlash control by Q2 emitter resistor, (b) trigger d.c. level set by RV, capacitive input, and transformer coupled output

the input trigger pulse can be a.c.-coupled *via* a capacitor, C, shown dotted in Fig. 7.5 (b). The output may also be a.c.-coupled by a transformer as also shown in Fig. 7.5 (b), and this gives an output pulse rather than a step voltage for an input step.

Practical Schmitt triggers

Three typical practical examples of Schmitt triggers are given in Fig. 7.6. In these it will be noted that the various component values do not agree exactly with the simplified design given earlier, but they will be found to be of the same order of magnitude, and the Schmitt is a delightful circuit in that many of its values are non-critical.

Fig. 7.6 (a) shows a conventional circuit to give 10 V output with a d.c. trigger level adjustment that enables the circuit to be driven by a 0·5 V p.-p. input up to 300 kHz.

Fig. 7.6 (b) shows a more sophisticated modification of the basic circuit to achieve operation from 100 Hz to 10 MHz. Note the

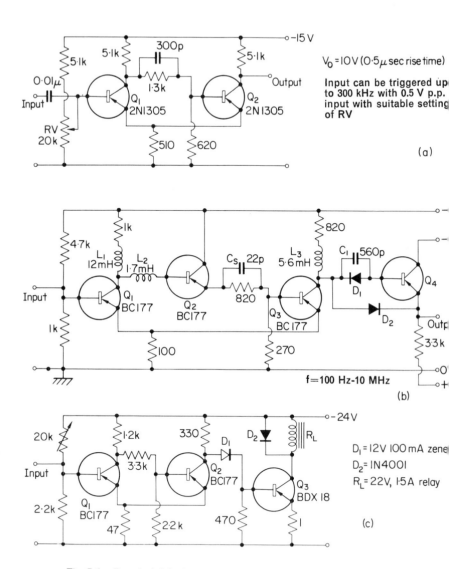

Fig. 7.6. *Practical Schmitt trigger circuits;* (a) *300 kHz 10 V output,* (b) *10 MHz Schmitt,* (c) *22 V, 1·5 A relay driver*

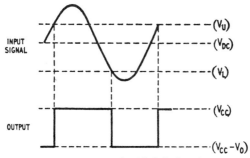

Fig. 7.7. 'Squaring' with Schmitt trigger

peaking coils L_1, L_2, L_3 inserted to enhance the high-frequency response. The emitter follower Q2 is used so that the speed-up capacitor, C_s, may be large without unduly increasing its circuit time constant. The output is taken from the collector of Q3 through the diode-capacitor filter, D1, C_1, to the base of the final emitter follower, Q4, which provides a low output impedance. D2 is a silicon diode used to protect the base-emitter diode of Q4 (which in a diffusion transistor of the type specified has a low voltage rating).

Finally, in Fig. 7.6 (c) we have a Schmitt trigger used to drive a 22 V, 1·5 A power relay. In this circuit, as only slow repetition rates are possible due to the speed limitation of the output alloy power transistor, the speed-up capacitor is omitted. The Zener diode D1 forms a useful d.c. coupling from the Schmitt output Q2 to the relay driver Q3. When the collector voltage of Q2 falls below 12 V, the Zener diode cuts off and isolates the output stage, so that it is not affected by the residual voltage on the Q2 collector. As a result, the transistor Q3 is able to cut off completely. The diode D2 is a clamp to short circuit any possibly harmful reverse voltage inductive spikes on the collector of Q3 when it switches off.

Applications

In general, the Schmitt provides a snap-action switch, and may simply be used as a fast-acting on-off switch.

The commonest application of the Schmitt, however, is probably as a *'squarer'*. This is illustrated in Fig. 7.7. Here an input sinusoidal signal shown at the top swings above the upper trip level, V_U and below the lower trip level, V_L. The output switches up to V_{CC} when the input rises above V_U and falls to $V_{CC}-V_U$, when the input falls below V_L. This gives a rectangular output for a sinusoidal input, i.e. it 'squares' the input signal. By adjusting the means input level, V_{DC}, halfway between V_U and V_L, we can make the output a true square wave with equal on and off periods. In this application, the Schmitt is really a *pulse-shaper*, converting sine waves into square.

If the input is a train of nominally rectangular pulses, whose shape has been degenerated, the Schmitt can be used as a *pulse reshaper or restorer*, converting the ragged input pulses into precise rectangular output pulses. Incidentally, since the output pulse amplitude is independent of the input pulse amplitude, the Schmitt trigger can also make a useful *pulse amplifier*. Again, because the output levels are only indirectly related to the input, the Schmitt can be used as a *signal level shifter*.

The Schmitt trigger often finds use as a *d.c. level detector*, since it can indicate positively when an input voltage rises above a specified reference level. When the reference level is zero volts, the circuit becomes a *zero cross-over detector*. Many more refined applications are possible. A particularly interesting one is where the input consists of an adjustable control level voltage mixed with a periodic time-varying signal, and the circuit is used as a *variable duty cycle switch* for certain control and regulator problems (*and possibly as a variable time delay device*).

CONCLUSION

An attempt has been made to outline the main design features and uses of the diode (or transistor) pump and the Schmitt trigger. With the ever-increasing use of digital pulse circuits in electronics, these are both becoming widely used as 'bricks' in switching systems. They merit fuller exposition than is possible in the scope of this chapter. Unfortunately no current textbooks give completely satisfactory analysis of these two circuits in their semiconductor versions. Until such become available, this chapter should at least give some general guidance as to their design and applications.

CHAPTER 8

Blocking Oscillators

In modern electronics there is often a requirement for a large-amplitude, sharp-edged pulse of short duration (anything from a millisecond down to a fraction of a microsecond). One circuit widely used for this purpose is the blocking oscillator—particularly where the pulses have to be widely spaced in time. The blocking oscillator is basically a single-transistor, transformer-coupled-feedback oscillator with special properties. Anyone interested in electronics should know something of its design and uses.

Practical blocking oscillator circuits, stripped of refinements, reduce to one or other of the three basic feedback arrangements of Fig. 8.1. In Fig. 8.1 (a) feedback is from collector to base with phase reversal through the transformer; in Fig. 8.1 (b) from collector to emitter in phase; and in Fig. 8.1 (c) from emitter to base in phase. Whatever configuration is used, the circuit produces an output pulse with the characteristic general shape shown in each case in Fig. 8.1, i.e., a rectangular pulse with a reverse overshoot on the trailing edge.

The blocking oscillator may be arranged as a monostable ('single-shot') circuit which remains switched off until suitably triggered and then produces a single output pulse. Again it may be arranged as an astable ('free-running') circuit which produces by itself a string of pulses at a fixed repetition rate without the need for external triggering. The principal difference between astable and monostable versions is that the free-running version has a forward d.c. bias on the emitter and the monostable a reverse bias. Apart from this, design considerations are largely the same for both versions.

While single pulses can be provided only by a triggered mono-

114 *Elements of Transistor Pulse Circuits*

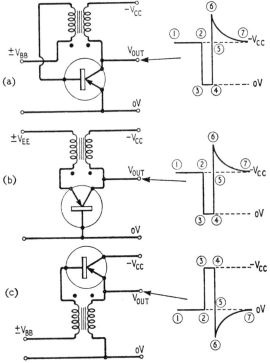

Fig. 8.1. Three basic feedback arrangements in blocking oscillator; (a) *collector-base,* (b) *collector-emitter,* (c) *emitter-base*

stable blocking oscillator, a pulse train can be provided in three ways : from (a) an astable blocking oscillator free running by itself, or (b) an astable blocking oscillator synchronised by an input pulse train at a frequency higher than the natural repetition rate of the astable, or (c) a monostable suitably triggered by an input pulse train.

The blocking oscillator circuit was apparently first developed by Appleton, Herd and Watson Watt (British Patent 23254) about 1923. When transistors arrived on the scene in 1948, it was soon found that the circuit took kindly to them. Whether with valves or transistors, however, the rigorous theoretical design of the blocking oscillator is extremely complex. On the other hand, the basic working principle is not all that difficult to understand, and it is possible to arrive at a simplified practical approximate design procedure. With this you can build a blocking oscillator to give something near the results you want, and then adjust to your exact requirements by intelligent cut-and-try at the end.

OPERATING PRINCIPLE

Ignoring for the present the difference between astable and monostable versions, let us consider generally how a blocking oscillator works. Fig. 8.1 (a) can be used for illustration. Assume for a start that the base bias voltage, V_{BB}, is zero. As the base-emitter diode is not forward biased, the transistor is cut off (no collector current flowing). In the absence of collector current, the collector is at the negative rail potential, point (1) on the output waveform illustration. If now a short negative pulse voltage, $-V_{BB}$, is applied to the transistor base (via the transformer winding), the base-emitter diode will go into forward conduction and collector current will begin to flow. This will cause the collector potential to start moving from the negative rail towards zero volts, from point (2) on the waveform. This positive-going change in collector voltage is phase inverted through the transformer and appears as a negative-going voltage at the transistor base. Phasing dots are inserted in Fig. 8.1 (a) to make this easier to follow. The negative-going feedback voltage at the transistor base adds regeneratively to the negative trigger input voltage, and the transistor switches very rapidly into bottoming. This means that the collector voltage approaches close to zero volts at point (3) on the waveform.

The full rail voltage is now impressed across the transformer collector winding. The transistor remains bottomed during the pulse period, from point (3) to (4), and the current through the transformer builds up linearly until it is limited by some circuit parameter.

When the transformer current stops increasing at point (4), the feedback voltage to the base disappears, and the transistor begins to switch off. This causes the collector current to fall off, and the collector voltage to move back from zero volts towards the negative voltage rail again. This negative-going collector voltage is reflected by phase inversion through the transformer, as a positive-going voltage at the transistor base. This tends to turn the transistor farther off (provided, of course, that the input trigger pulse, $-V_{BB}$, has ceased). The circuit thus switches regeneratively back to the condition where the transistor is completely cut off, and the collector is once more at the negative rail voltage, point (5) on the waveform.

116 *Elements of Transistor Pulse Circuits*

Because of transformer inductance, however, the collector voltage overshoots the rail voltage to reach point (6) before returning finally to the rail voltage, $-V_{CC}$ at point (7). This end point corresponds to the initial state at point (1), and the blocking oscillator is ready to be triggered again.

The exact shape of the overshoot depends on the various losses and reactance in the practical circuit, but, other things being equal, the height of the overshoot tends to be proportional to the pulse width. Whatever the exact pulse shape, all blocking oscillator output pulses have this 'trade mark' of an essentially rectangular pulse with a reverse 'pip' on the back end.

TIMING MECHANISMS

We mentioned above that the pulse terminates when circuit limitations prevent further increase in the magnetising current in the transformer. In practice, circuit limiting methods used reduce to six as follows:

(1) *'Beta-limitation'*, where the collector current limits in the end because the transistor current gain ('beta') has a finite upper limit.

(2) *'Core-saturation'* where, the transformer core is designed to saturate when the magnetising current rises sufficiently to bring the core magnetic flux up to its limiting value.

(3) *'RC-limitation'*, where the standing d.c. bias on the circuit is set up by an RC network, and the 'on' base current is finally limited by the discharge of the capacitor.

(4) *'Switch-off-limitation'*, where the pulse is precisely terminated by an external reverse trigger pulse applied to an input.

(5) *'Tuned-circuit-limitation'*, where a tuned circuit is inserted in the feedback loop. This oscillates when pulsed and gives an initial peak to start the blocking cycle and a second peak separated by a defined time to terminate it.

(6) *'Delay-line-limitation'*, where a pulse generated by the blocking oscillator switch-on is fed into a delay line, and the pulse coming out of the delay line after a fixed delay switches the circuit off.

Which of these timing mechanisms you use depends on several factors. Where very precise pulse width control is essential, it is advisable to use either switch-off, tuned-circuit, or delay-line limitation, because they give pulse widths virtually independent of supply voltage, circuit loading and transistor characteristics (except minority carrier storage). However, it can be quite expensive to ensure precise pulse widths by these methods. The other control methods may give less precise pulse widths but are more economical. Limiting by the transformer inductance (as in beta or core limiting) or by C (as in RC limiting) is simple and cheap. With beta or RC limiting, the pulse width is sensitive to circuit loading and transistor characteristics—particularly input impedance. With core saturation, the pulse width is less dependent on transistor characteristics

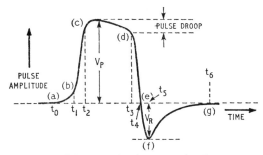

Fig. 8.2. Details of output pulse shape

and circuit loading; but in this case pulse widths are sensitive to supply voltage and consideration may have to be given to core-resetting.

SWITCHING CYCLE DETAIL

We showed idealised vertical edges to the output voltage pulse in Fig. 8.1 earlier. A real pulse would take a shape more like Fig. 8.2. Here the pulse is initiated at point (a), time t_0. At point (b), time t_1, it has risen to 10% of the full amplitude and $(t_1 - t_0)$ is known as the 'delay time'. At point (c), time t_2, the pulse has risen to 90% of the full amplitude, V_P, and $(t_2 - t_1)$ is known as the 'rise time'. The pulse top, ideally flat, actually 'droops' to point (d), time t_3, before the circuit begins to switch off. The pulse width can be taken as $(t_3 - t_2)$. The pulse then falls to zero at point (e), time t_4, and $(t_4 - t_3)$ can be taken as the 'fall time'. The pulse overshoots to a reverse amplitude V_R, at point (f), time t_5, before returning to zero at point (g), time t_6. Here $(t_6 - t_4)$ is the blocking oscillator 'recovery time'.

For pulse lengths greater than a few microseconds, the delay, rise and fall times are so short with modern transistors that they can be ignored in most designs, provided a suitable transistor is selected. Complex formulae exist for computing these rapid transient switching times, but for 'ordinary' requirements they are of somewhat academic interest. A practical rule of thumb is to select a transistor with f_{co} greater than $(10/T_P)$ MHz where T_P is the desired pulse length in microseconds. The pulse rise and fall times will then be sufficiently fast compared with the pulse duration to be ignored. (For very short pulse widths of the order of a microsecond or less, it may not be possible to ignore the switching times and the designer will then have to consult one of the standard full treatments such as 'Junction Transistor Blocking Oscillators', by J. G. Linvill and R. H. Mattson, *Proc. I.R.E.*, Vol. 43, No. 11, November 1955, pp. 1632–1639.) For the rest of this chapter we will assume negligible rise and fall times.

Pulse droop too is not a critical problem with most designs, and except for very long pulse lengths of the order of several hundred microseconds can be ignored in practice. We will not deal further with it in this chapter.

DESIGN REQUIREMENTS

Ignoring rise and fall times and pulse droop, we now turn to the main basic design requirements for a blocking oscillator. Anyone trying to design a blocking oscillator has to meet requirements of (a) pulse width, (b) pulse amplitude, (c) load resistance, (d) overshoot amplitude, and (e) recovery time.

The basic circuit of a blocking oscillator has three degrees of freedom:

(a) *The timing mechanism* may be of any the several methods detailed earlier. It is not possible to give here a satisfactory treatment of all of these. It is proposed to limit the discussion to beta and core limiting circuits. Readers interested in RC, switch-off, tuned-circuit, or delay-line termination of the pulse should consult standard text-books such as S. Schwartz *Selected Semiconductor Circuits* (Wiley, 1960).

(b) *Saturated or non-saturated transistor* operation may be used. When we discussed the operating principle earlier, we assumed the transistor saturated or 'bottomed' during the pulse, and this circuit is commonly referred to as a 'saturated' blocking oscillator. For higher speed operation, it is sometimes desirable to prevent the transistor bottoming by clamping its collector through a diode to a fixed 'hold-off' voltage source. We will consider the simpler

Blocking Oscillators 119

saturated blocking oscillator first and cover non-saturated versions later.

(c) *The transformer feedback mode* must be chosen appropriately, from collector to base, collector to emitter or emitter to base.

ILLUSTRATIVE DESIGN FOR MONOSTABLE SATURATED BLOCKING OSCILLATOR

As a first approach to the design, we will consider a monostable, saturated, collector-base feedback, blocking oscillator and give a 'cookery-book' design equation for both beta and core limiting which are simple and accurate enough for most applications. Fig. 8.3 (a) gives the basic circuit. The circuit as shown can be either beta or core limited. Both timing mechanisms use the same type of circuit (although component values will not be the same).

Theoretically the base and emitter resistors, R_B and R_E, are not essential to the operation. However, external resistances are normally included in practical oscillators to swamp the effect of the internal resistances of the transistor.

The load resistor R_O is shown capacitively coupled, but other ways of connecting the load resistance are discussed later.

When you are designing a blocking oscillator, you are aiming in the first place for a pulse of specified width, T_P, and amplitude, V_P, feeding into a load, R_O. The most important feature of the design is usually the pulse width, and this is determined primarily by the transformer characteristics.

In the collector-base feedback circuit of Fig. 8.3 (a), *if it is beta limited*, the inductance of the transformer collector winding, L_C, necessary to give a pulse width T_P is given approximately by:

$$L_C \text{ (beta limited)} = \frac{T_P}{\frac{N_B/N_C}{R_B + R_E} - \frac{1}{R_O}} \qquad (8.1)$$

where N_B and N_C are the number of turns in the base and collector windings of the transformer, R_B and R_E are the external base and emitter resistors, and R_O is the load resistance.

To give a pulse height, V_P, the supply rail voltage, V_{CC}, required is given by:

$$V_{CC} \text{ (beta limited)} = V_P \left(1 + \frac{R_E N_B/N_C}{R_B + R_E}\right) \qquad (8.2)$$

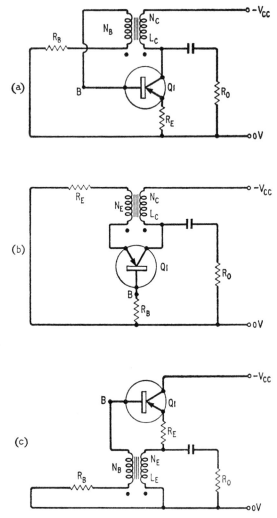

Fig. 8.3. Saturated monostable blocking oscillator design circuits; (a) collector-base feedback, (b) collector-emitter feedback, (c) emitter-base feedback

These formulae assume that a transistor with a reasonably high beta is used—say better than about 30.

To ensure that the blocking oscillator is actually operating in the beta limiting mode, the transformer must be such that its core does not saturate under the peak magnetising current in the windings. It can be shown that the peak magnetising current is given approximately by:

$$I_M \text{ (beta limited)} = V_P \left(\frac{N_B/N_C}{R_B + R_E} - \frac{1}{R_O} \right) \qquad (8.3)$$

Before we can use equations (8.1) to (8.3), we must select tentative design values for N_B/N_C, R_B and R_E.

The transformer turns ratio, N_B/N_C, is governed by several factors. To give enough loop gain for the circuit to block on the pulse, it can be shown that N_B/N_C must be much less than h_{FE}, the d.c. current gain of the transistor. For fast switching between on and off, values of N_B/N_C between 1/5 and 1/1 are common. As h_{FE} is usually higher than 10, a turns ratio of this order will adequately ensure blocking. An empirical value of 1/5 is commonly used by designers.

R_B and R_E are usually selected at not less than ten times the related transistor internal resistances. With typical modern transistors this leads to lower limit values of about 500 ohms for R_B and 20 ohms for R_E.

To get the design under way, you can use the value of turns ratio and external resistances outlined above. Equations (8.1) to (8.3) then give a first estimate of suitable L_C, V_{CC} and I_M. For the value of L_C thus computed, select a suitable core and number of collector winding turns, N_C. Confirm that $N_C I_M$ is well within the ampere-turns saturation limit of the core. Select a transistor with f_{co} greater than $10/T_P$, with a voltage rating greater than the computed V_{CC}, and with a base current rating of not less than I_M. Several attempts at suitable values of R_B and R_E may be necessary before satisfactory results are achieved. Finally, make up the circuit, check T_P and V_P on an oscilloscope, and marginally adjust R_B and R_E to give the exact results required. For any given transformer, quite a wide range of variations of T_P and V_P can be effected by suitably varying these two resistor values.

So far we have been dealing with the beta limited version of the circuit of Fig. 8.3 (a). Although it does not show up in our approximate formulae, both the pulse width and height are partially

122 *Elements of Transistor Pulse Circuits*

dependent on the transistor parameters in this case. Where it is desired to reduce the effect of transistor variations from the design, it is common to design for the core saturated mode of the blocking oscillator.

For transformer core saturation operation of the circuit in Fig. 8.3 (a), the formulae become:

$$L_C \text{ (core limiting)} = \frac{N_C V_P T_P}{H_K} \tag{8.4}$$

where H_K = limiting coercive force of the core selected.

$$V_{CC} \text{ (core limiting)} = V_P \left(1 + \frac{R_E N_B/N_C}{R_B + R_E}\right) \tag{8.5}$$

$$I_M \text{ (core limiting)} = H_K/N_C \tag{8.6}$$

With the same sort of cut-and-try experimentation as before, it is possible to arrive at a suitable design of the transformer for this case.

These design procedures may seem a little vague, but they are really quite practicable, and put you into the right area of operation to make final small adjustments to get your exact pulse requirements. In practice, incidentally, you can easily tell whether a circuit is beta or core limiting. If you vary V_{CC} and find T_P also varies materially, transformer core limiting is indicated. Otherwise the circuit is beta limiting.

SATURATED BLOCKING OSCILLATORS WITH COLLECTOR-EMITTER AND EMITTER-BASE FEEDBACK

Figs. 8.3 (b) and 8.3 (c) are the collector-emitter and emitter-base feedback versions of the collector-base feedback circuit discussed in (a). Equivalent simplified formulae for these new versions are:

(a) Saturated, collector-emitter feedback—Fig. 8.3 (b)

(1) Beta limited:

$$L_C = \frac{T_P}{\dfrac{N_E/N_C}{R_B + R_E} - \dfrac{1}{R_O}}$$

$$V_{CC} = V_P \left(1 + \frac{R_B N_E/N_C}{R_B + R_E}\right)$$

$$I_M = V_P \left(\frac{N_E/N_C}{R_B + R_E} - \frac{1}{R_O} \right)$$

(2) Core limited:

$$L_C = N_C V_P T_P / H_K$$

$$V_{CC} = V_P \left(1 + \frac{R_B N_E/N_C}{R_B + R_E} \right)$$

$$I_M = H_K / N_C$$

In these cases, R_B and R_E are governed by the same considerations as before, and the optimum transformer turns ratio N_E/N_C also lies between 1/5 and 1/1.

(b) Saturated, emitter-base feedback—Fig. 8.3 (c).

(1) Beta limited:

$$L_C = \frac{T_P}{\dfrac{N_B/N_E}{R_B + R_E} - \dfrac{1}{R_O}}$$

$$V_{CC} = V_P \left(1 + \frac{R_E N_B/N_E}{R_B + R_E} \right)$$

$$I_M = V_P \left(\frac{N_B/N_E}{R_B + R_E} - \frac{1}{R_O} \right)$$

(2) Core limited:

$$L_C = N_E V_P T_P / H_K$$

$$V_{CC} = V_P \left(1 + \frac{R_E N_B/N_E}{R_B + R_E} \right)$$

$$I_M = H_K / N_E$$

In these cases, the optimum transformer turns ratio, N_B/N_E, lies between 1·2/1 and 2/1. It is common to start with a value of 1·2/1.

NON-SATURATED BLOCKING OSCILLATORS

Where very short pulses are required, a common circuit device is to diode-clamp the transistor so that the collector-emitter voltage during the pulse cannot fall below a certain limit. This gives rise to the family of non-saturated blocking oscillators shown in basic form in Fig. 8.4. The three possible feedback modes are illustrated.

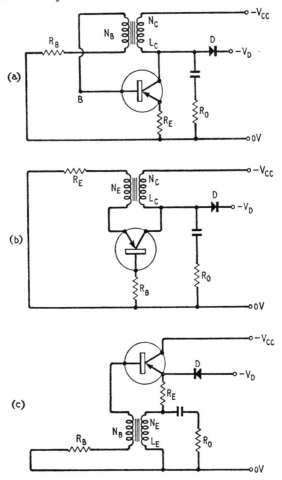

Fig. 8.4. Non-saturated monostable blocking oscillator circuits; (a) collector-base feedback, (b) collector-emitter feedback, (c) emitter-base feedback

In all the cases of Fig. 8.4 the transformer turns ratio and external resistor values are set by the same considerations as apply in the saturated unclamped blocking oscillator cases discussed earlier. Those interested in this type of circuit should consult a standard treatment such as given in H. J. Reich *Functional Circuits and Oscillators* (Van Nostrand, 1961).

RECOVERY TIME

It can be shown that the recovery time of a blocking oscillator is of the order of $4L/R$ where L is the inductance of the transformer primary winding and R is the resistance reflected across that winding from the load and other circuit resistances. This sets an upper limit to L for a given load resistance when a maximum pulse repetition rate is prescribed, i.e. a minimum recovery time.

REVERSE SPIKE

The reverse spike on the back edge of the blocking oscillator pulse is basic to the circuit. In general the amplitude, shape, and width

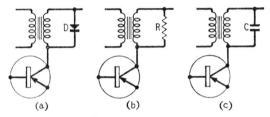

Fig. 8.5. Switch-off overshoot control circuits. Primary winding shunted by (a) diode, (b) resistor, (c) capacitor

of this spike depend on the various losses and reactances in the circuit.

It can be shown that the less the reverse spike is suppressed the more rapidly does the circuit recover ready to receive the next trigger pulse. A common design approach is to fix all the various circuit parameters in such a way that the spike height is within the collector-emitter voltage rating of the transistor, but not to damp it too severely. This avoids excessively long recovery times.

The ways commonly used to control the shape of the spike are illustrated in Fig. 8.5. The most usual technique is to connect a diode across one of the windings, e.g. the collector as in (a). This diode clamps the collector voltage to the negative supply rail on switch off and almost completely removes the overshoot spike. Often a resistance is also placed in series with this diode to reduce the diode damping. Again a resistance (which may be merely the load resistance or some other circuit damping resistance) is placed across the transformer winding and the diode dispensed with as shown in Fig. 8.5 (b). A final method sometimes used is to shunt the collector winding with a small capacitance as in Fig. 8.5 (c). It is emphasised that some form of damping must be used in practice, otherwise the blocking oscillator may go into oscillation or, if not, may 'ring' so that the end of the pulse degenerates into a damped series of sine waves.

ASTABLE OPERATION

So far we have been considering only the monostable version of the blocking oscillator where the circuit goes into a blocked pulse only when it is actively triggered by some trigger pulse input. In practice the blocking oscillator is often required in a free-running version. For this the transistor is d.c. biased so that, if the circuit were not oscillating, the transistor would be conducting. This can be achieved in several ways. In Fig. 8.6 (a) the resistor R forward biases the transistor. When the transistor blocks and gives out a pulse, it also discharges the capacitor C_B. At the end of the pulse C_B begins to recharge through R until once again the transistor is forward-biased and blocks once more. This circuit will then block and relax, block and relax, with a pulse repetition rate set primarily by the time constant RC_B.

An alternative biasing method is shown in Fig. 8.6 (b) where the repetition rate controlling network, RC_B, is a parallel one. In this case the capacitor charges when the circuit blocks and then in the relaxation period discharges through R. Here once again the repetition rate is set largely by the time constant RC_B.

In Figs. 8.6 (c) and (d), the time constant for pulse repetition is set in the emitter circuit by a series or parallel arrangement $R_E C_E$. Here again the repetition rate is proportional to the $R_E C_E$ product.

With astable circuits of this type the main problem is to provide a pulse of known width at a known and preferably variable pulse repetition rate. Unfortunately in practice it will be found that when you try to vary the pulse repetition rate by varying R you will at the same time vary the pulse width T_P. Circuits have been derived, however, where the pulse repetition rate can be made independent of the pulse width, e.g. Fig. 8.9 (b).

TRIGGERING

Where a blocking oscillator is controlled by a trigger pulse, this trigger can be applied in a variety of ways. A common one is to apply a sharp negative-going pulse to the base via an isolating capacitance such as C_T in Fig. 8.7 (a). The capacitively coupled

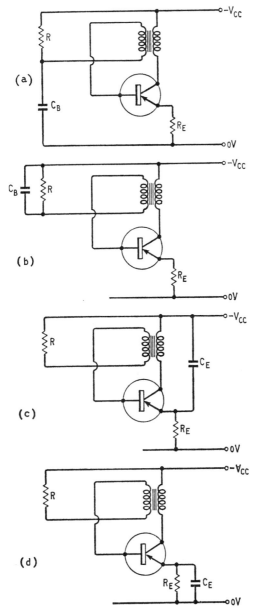

Fig. 8.6. Astable blocking oscillator RC bias networks;
(a) base, series, (b) base, parallel, (c) emitter, series,
(d) emitter, parallel

128 *Elements of Transistor Pulse Circuits*

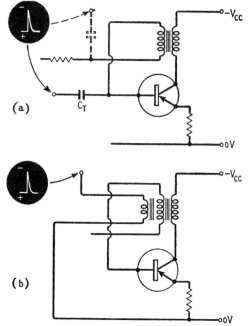

Fig. 8.7. Triggering circuits; (a) via capacitor, (b) via transformer winding

trigger could also be fed into the other end of the base winding, as shown dotted in the same illustration.

Although capacitor coupling of the trigger pulse is by far the commonest, in some cases the trigger is supplied via a separate transformer winding as shown in Fig. 8.7 (b). Obviously this method adds complexity to the transformer specification, and is normally avoided.

The requirements on the trigger pulse are not very stringent in blocking oscillators. The most obvious one of course is that the width of the trigger pulse should be less than that of the blocking oscillator output pulse, otherwise the switch-off at the end of the output pulse can be complicated. On the other hand the trigger pulse must not be too short, otherwise due to the slow switching speed in the transistor and the effect of stray circuit capacitances the switch-on action may not be firmly initiated. Here a little experimentation with the trigger shape may be necessary to arrive at an optimum trigger amplitude and width. As a rough rule of thumb, some designers aim for a trigger pulse about 20% of the minimum output pulse width.

OUTPUT TAKE-OFF

The pulse can be taken off from a blocking oscillator either directly, capacitively or inductively. In Fig. 8.8 (a) the output load R_O is shown directly in series with the transistor. A common alternative is Fig. 8.8 (b) where the output is taken capacitively from the transistor collector to the output load R_O, or as shown dotted in the same circuit from the emitter resistor, R_E.

A final method is shown in Fig. 8.8 (c) where a tertiary winding in the transformer is used to couple to the load. This last method has a big advantage that the polarity of the output pulse can be selected at will.

As can be seen from the earlier analysis, the load resistance can have quite a significant effect on pulse output width and height. Where a blocking oscillator is to be used with a variety of different loads, it is common to use a buffer emitter-follower so that the varying output loads have a relatively negligible effect on the pulse characteristics.

APPLICATIONS

The primary purpose of a blocking oscillator is to generate pulses of large peak power in a train of low mean power. For example, it is possible to use an r.f. alloy transistor of 100 mA mean current rating to give a peak pulse of 2 amps. The average power dissipation is kept within the transistor mean power rating since the duty cycle is low. This property of the blocking oscillator is particularly useful for the rapid discharge of a capacitor. In this fashion the blocking oscillator is commonly used for sawtooth wave generation. A full discussion of this will be found in the chapters on line-time base and field-time base circuits in T. D. Towers' *Transistor Television Receivers* (Iliffe, 1963).

The blocking oscillator is widely used as a very low impedance switch. Typical of this is when it is used to drive a silicon controlled rectifier, or is used as a voltage comparator switch.

One use of the blocking oscillator is as a source of pulses of defined shape. A specific case of this is where a blocking oscillator triggered from a high-stability crystal oscillator is used as a clock pulse source to synchronise the various functions of a computer.

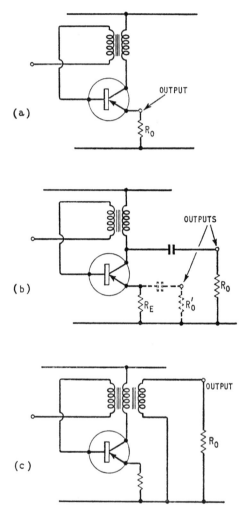

Fig. 8.8. Output take-off; (a) *via series resistance,* (b) *via capacitor,* (c) *via transformer winding*

Pulse reshaping, i.e. converting ragged pulses back to a standard shape, is another common use of the blocking oscillator.

An astable blocking oscillator can be used as frequency divider by synchronising it with an input signal of a frequency several times its own natural repetition rate. In this way it has been widely used in divider type electronic organs, where the control frequency is slightly faster than twice the natural frequency of the blocking oscillator being triggered. By this means, each of the blocking oscillators in a divider chain operates at a frequency one half of that immediately above it.

In Fig. 8.9 will be found three typical practical examples of blocking oscillator circuits. The circuits have been given in sufficient detail for the interested constructor to make them up.

In Fig. 8.9 (a) is shown a simple 10 μsec-pulse-width, monostable, collector-base feedback, saturated blocking oscillator. From the details on the circuit diagram it will be noted that a 5 : 1 turns ratio is used in the transformer. To show how the practical simplified formulae do in fact work, it should be noted that this circuit was computed to work at a pulse width of approximately 11 μsec and turned in the event to be 10 μsec. By adjusting the slug of the ferrite core of the transformer it was possible to vary the pulse width by $\pm$ 5% on the actual value found.

The circuit given in Fig. 8.9 (b) is that of a 1 μsec astable blocking oscillator beta limited and non-saturated. The circuit is such that it is possible to vary the p.r.f. from 50 to 250 kHz by varying the positive control voltage on the emitter of the blocking oscillator transistor and yet keep the pulse width materially constant. This is possible because of the Miller-effect 1000 pF capacitance from the collector to the base of the other transistor. The 1000 pF capacitor across the emitter of the blocking oscillator transistor is designed to sharpen the front and back edges of the pulse. The collector of the blocking oscillator transistor is diode-clamped by a three-turn overwind on the collector winding which prevents the transistor going into bottoming. The diode and the 1 kΩ resistance in series across the total collector winding is the normal reverse spike suppression circuit. The 1 kΩ value is relatively high and places a minimum damping on the transformer circuit to ensure that the recovery time is not degraded.

Fig. 8.9 (c) shows a typical s.c.r.-firing monostable providing a 200 μsec, 16 V output pulse in a collector-emitter feedback, saturated transistor, arrangement. The diodes D1, D2, D3 are inserted to provide satisfactory thermal stability at high ambient temperatures

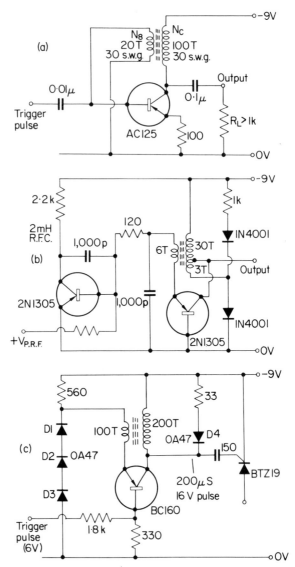

Fig. 8.9. *Practical examples.* (a) *Simple 10 μsec monostable, collector-base feedback. Transformer on LA2517 pot core;* (b) *1 μsec astable with p.r.f. variable from 50 to 250 kHz. Transformer on FX 1011. P.r.f. 50–250 kHz or* $V_{p.r.f.} = 2\text{--}12\ V$; (c) *s.c.r.-firing monostable with 200 μsec, 16 V output pulse. Transformer on FX 1238 (X2)*

Blocking Oscillators 133

up to 45°C. This circuit is suitable for driving a typical s.c.r. such as the BTZ19.

CONCLUSION

In this chapter on blocking oscillators I have attempted to give the reader some idea of the main factors which govern the design of the blocking oscillators and even to enable him to make up a few circuits to try his hand. In practice it will be found that if you use variable resistors for R_B and R_E, you can 'wind the circuits up and down' and test out the effects of varying the controlling circuit elements. The references quoted at various points in the chapter should serve as a guide to those interested in a more rigorous approach to the design of blocking oscillators.

CHAPTER 9

'Gates'

You can look at electronics as simply 'piping' a flow of signals through a system. Electronic 'gates' are the circuits by which you steer these signals the way you want them to go. In the 'old days', before the rise of the electronic computer, gating circuits were not very important in the framework of electronics and were seldom given an extended treatment in the textbooks. Nowadays, with pulse circuitry playing a very large role in electronics, gates have become an important special field with its own special jargon. The purpose of this chapter is to try to present a working overall picture of this field, insofar at least as it is concerned with transistors and semiconductor diodes.

TRANSMISSION V. LOGICAL GATES

There are two basic kinds of gates, 'transmission' and 'logical'. A transmission gate is one in which, ideally, the output is identical with the input signal when the gate is open, and there is no output when the gate is closed. The transmission gate is sometimes known as a 'linear' gate for this reason—the output signal, when there is one, is the same as the input. In a logical gate on the other hand, the output may or may not be a replica of the input. What actually appears at the output is decided by the logical combination of the signals fed into the gate inputs. The output signal, when there is one, need not be the same as any of the input signals.

A commonplace illustration of a transmission gate is the on/off switch of a mains radio receiver. Here the input to the gate is the 230 V mains voltage. When the switch is off, the switch output to

'Gates' 135

the receiver is zero; when it is 'on', the output is 230 V. An illustration of a logical gate from the same receiver is the a.g.c.-controlled 1st i.f. stage. Here the 470 kHz signal from the mixer and the a.g.c. d.c.-control voltage from the detector are fed into the controlled stage and the output is determined by the combination of these two.

TRANSMISSION GATES USING DIODES

Diodes are often used for building transmission gates, because they approximate to a short circuit when forward-biased and to an open circuit when reverse-biased. A simple illustration is the unidirectional *single-diode gate* in Fig. 9.1 (a). Here, when the control voltage V_G applied to the gate control terminal is positive, the diode D conducts and the signal input $+V_S$ reappears as $+V_O$ at the output. When the gate control voltage V_G is negative the diode is reverse biased and no signal appears at the output. The circuit is thus equivalent to the simple on-off switch of Fig. 9.1 (b) controlled by V_G.

The simple circuit of Fig. 9.1 (a) has the defect that the output signal V_O when the gate is open includes not only the input signal V_S but a d.c. 'pedestal' voltage due to the forward gate current through the diode into the load R_O. Also some of the gate current is deflected through R_G into the signal source, so that the signal line and the gate-control circuit are not adequately isolated.

To deal first with the spurious pedestal voltage at the output, this can be eliminated by applying a balanced gate control as in the *two-diode gate* of Fig. 9.1 (c). Here a push-pull gate voltage $\pm V_G$ is applied across diodes D1, D2, which are so arranged that, when V_G is positive, both diodes conduct and when V_G is negative both are cut off. So far as the gate signal goes, the diodes are in series, but, so far as the controlled signal V_S is concerned, the diodes are in parallel. The important point is that the voltages at points (a) and (b) are not affected by the application of the push-pull gate voltages.

The circuit of Fig. 9.1 (c) has certain defects still however. The 'gain' of the gate (defined as V_O/V_G during transmission) can be shown to be

$$A = \frac{R_G}{R_G + R_S} \cdot \frac{2R_O}{2R_O + R_G R_S/(R_G + R_S)} \qquad (9.1)$$

if we ignore the forward diode resistances. This points to the major

136 *Elements of Transistor Pulse Circuits*

disadvantage of the two-diode gate—its low 'gain' i.e. heavy attenuation of the controlled signal voltage. It is also susceptible to unbalance in the gating voltages.

Better gain can be achieved by the *four-diode circuit* of Fig. 9.1 (d) which requires two additional fixed bias voltages V_K and $-V_K$. As these are fixed, however, they do not unduly complicate the circuit. The diodes D1 and D2, considered by themselves, would be forward-biased by the voltages $+V_K$, $-V_K$. Now if V_G is positive, diodes D3 and D4 are cut off, D1 and D2 remain conducting, and the gate transmits V_S to the output. If now the gate voltage V_G is made sufficiently negative, diodes D3 and D4 conduct, D1 and D2 cut off,

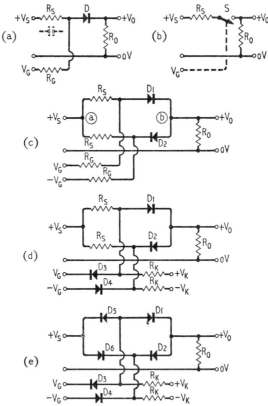

Fig. 9.1. *Transmission gates with series diodes; (a) basic single diode circuit, (b) equivalent single-diode circuit, (c) two-diode gate, (d) four-diode gate, (e) six-diode gate*

'Gates' 137

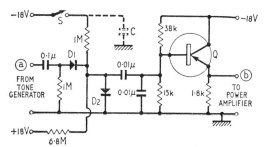

Fig. 9.2. Practical diode transmission gate used in electric organs

and the gate ceases to transmit. It can be shown that this four-diode gate has a higher gain that the two-diode version, but it is still far from the theoretical unity gain of a perfect transmission gate. However the circuit is now much less sensitive to gate voltage unbalance.

For further improvement in gain, recourse may be had to the *six-diode transmission gate* of Fig. 9.1 (e). Here the signal-isolating resistors R_S are replaced by diodes D5, D6 and a gain very close to unity can be achieved if R_K and R_O are large compared with the forward diode resistances.

We have analysed these diode transmission gates in terms of d.c. voltages, but equally they can be used with pulse or a.c. voltages. The coupling capacitors, where used, will then have values governed by the usual design considerations i.e. no significant discharge between pulses and no significant charging on pulses. Also account may have to be taken of diode capacitances and leakage currents. Fuller information can be found on these aspects in J. Millman and T. H. Puckett 'Accurate Linear Bidirectional Diode Gates' *Proc. I.R.E.* Vol. 43, pp. 27–37, Jan. 1955.

All the diode transmission gates of Fig. 9.1 work on the principle of short- or open-circuiting a diode in *series* with the signal line. A diode can similarly be used as a switch *shunting* the signal line. This gives rise to a whole family of shunt diode transmission gates which can be derived by analogy from Fig. 9.1.

Diode transmission gates find wide use in electronics. A typical example taken from the field of electronic organs, is given in Fig. 9.2 to illustrate actual circuit values used in practice. Here a series diode D1 is switched on and off by a switch S attached to the organ manual key. When the note is depressed, S closes and forward biases D1 by applying −18 V across the diode and the two series

1 MΩ resistors. At the same time diode D2 is reverse-biased and effectively open circuit. The tone signals arriving at (a) then pass through to the output (b) via a buffer emitter-follower stage Q. When the key is released, the switch S opens, and the -18 V is removed from the diode circuit. Diode D1 is now reverse-biased by the polarising $+18$ V through 6·8 MΩ, and presents an open circuit to the input signal. At the same time diode D2 becomes forward biased and clamps the signal line on the output side of D1 to earth. The two 0·01 μF capacitors and 15 kΩ resistor at the input of transistor Q form a filter network to partially suppress switching transients ('key clicks') on the signal line. The capacitor C shown dotted can be added to the circuit so that when the note is released (switch S opens) the polarising voltage holding D1 conducting dies away only gradually (with a time constant set by the value of C and its associated 1 MΩ resistor). As a result the note does not cut off sharply, but 'sustains' for a short time to simulate the gradual decay of the note from a conventional organ pipe.

TRANSMISSION GATES USING TRANSISTORS

The effectiveness of diode transmission gates is limited somewhat because, although the diode forward resistance is small, it is still far from the ideal of zero resistance. It has been found that with transistors you can get much lower forward resistances. There are two ways of using transistors in transmission gates.

First, the transistor can be substituted directly for a diode in any of the gate circuits explored earlier, if its collector and base are directly connected together in what has been called the 'tridode' (= triode-as-diode) arrangement shown in Fig. 9.3 (a) for p-n-p or (b) n-p-n. This produces a diode-substitute with a forward resistance several times lower than an equivalent conventional point contact or gold-bonded diode.

Secondly, the transistor can be operated as a triode, where the control gate voltage is applied to the base and the controlled signal voltage to the collector (or to the emitter used as a collector). A single-ended arrangement of this sort is given in Fig. 9.4 (a). Here a positive signal $+V_S$ applied to the input is not transmitted to the output if a positive gate voltage $+V_G$ is applied to the n-p-n transistor base. The transistor is then 'bottomed' and has a very low collector output resistance which in effect short-circuits the signal line to earth. If a negative gate voltage $-V_G$ is applied to the transistor base, the transistor has both emitter and collector junctions

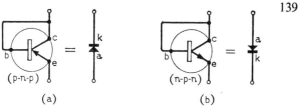

Fig. 9.3. Transistor 'triode' substitute for diode

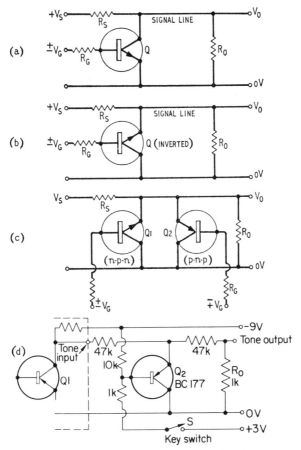

Fig. 9.4. Transistor transmission gates; (a) basic single-ended (positive unidirectional input), (b) basic inverted single-ended (positive unidirectional input), (c) basic push-pull (bidirectional input), (d) electronic organ keying gate

140 *Elements of Transistor Pulse Circuits*

cut off. Under these conditions the collector-emitter resistance of the transistor is high, the signal is not significantly shunted to earth and the input signal V_S is transmitted to the output as V_O with an attenuation fixed by the ratio of R_O to R_S.

If the switching transistor is inverted (i.e. with collector and emitter interchanged) as in Fig. 9.4 (b), it is found that the 'on' resistance may be much lower than for the non-inverted configuration, while the off resistance is much higher. The inverted configuration thus gives a closer approach to an ideal switch.

For negative-going input signals, the circuits of Fig. 9.4 (a) and (b) can be used, but with p-n-p transistors instead of n-p-n and with a negative gate voltage V_G to close the gate and a positive one to open it.

When the input signal V_S is bidirectional (i.e. either positive or negative), a balanced arrangement of two transistors such as Fig. 9.4 (c) may be used. Here the n-p-n transistor Q1 controls +ve signals while the p-n-p one, Q2, controls negative signals. Note that oppositely-phased gate sign signals V_G are used on the two transistors.

Some people will recognise in Fig. 9.4 (a) and (b) the basic circuit of the transistor 'chopper' widely used in d.c. amplifier practice. In this application the gate may be controlled by a square wave with a fixed repetition rate. This converts a d.c. input into a 'chopped' square wave output which can be handled by a standard a.c. amplifier and thus largely reduce the effect of drift which is so troublesome in a 'straight' direct-coupled amplifier.

An interesting practical use of this type of circuit is again in electronic organs as an alternative to the diode keying system described earlier. This is illustrated in Fig. 9.4 (d) where the output from a multivibrator tone generator Q1 is passed via a 47 kΩ isolating resistor to the top end of an inverted gating transistor Q2. The output of the tone generator is unidirectional negative so a p-n-p gating transistor is used. The switch S attached to the playing key is normally open (when the note is up), and the transistor Q2 is biased on by the 10 kΩ resistor from the base to the negative supply rail. The output resistance at the emitter of Q2 is then very low, and the signal line is effectively shorted to earth so that no tone signal appears at the output. When now the key is pressed, S closes and the base of Q2 is carried positive by the potentiometer formed by the 10 kΩ and 1 kΩ across the -9 V and $+3$ V rails. Transistor Q2 then has both its collector and emitter junctions reverse biased. Its emitter therefore presents a high impedance to the signal line.

'Gates' 141

Thus when a note is pressed, its tone is no longer shunted to earth and so appears at the output. By connecting various CR networks in the base circuit of the gate transistor it is possible to shape the envelope of a note to give various forms of attack and sustain.

Digital computers use transmission gates frequently, but they also require logical gates to carry out all their operations. Before the advent of transistors, most logical gates were constructed of diode networks. By themselves, however, diodes cannot make up a complete set of gates for a computer, since they cannot perform the

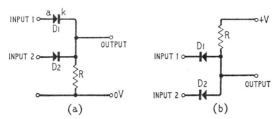

Fig. 9.5. Basic diode logic gates; (a) 'OR', (b) 'AND'

functions of amplification and phase inversion which must be available. With only diodes, however, we can make up at least two of the most widely used logical circuits—the 'OR' and 'AND' gates.

Before we examine the diode OR and AND gates, we must diverge to set the picture of logic 'levels'. A digital computer usually works by handling a series of pulses. When the pulse is there, the signal level can be said to be 'UP' and when it is absent 'DOWN'. For the purpose of this discussion, we will regard a signal as UP when the voltage level is $+V$ and DOWN when it is zero. In practice the UP and DOWN levels may be arbitrarily selected voltages, neither of them zero.

The basic circuit of a *diode OR gate* is given in Fig. 9.5 (a). When input 1 is UP (i.e. at $+V$) the diode D1 is forward biased because its anode (a) is positive with respect to its cathode (k) (which is connected via R to zero volts). Then the output, too, is at $+V$, if we assume an ideal diode which is a short circuit when forward biased. In the meantime, if input (2) is DOWN (at zero volts), diode D2 is reverse biased, and presents an effective open circuit across the output, not affecting the output voltage level. If, however, input (2) is also UP, both diodes will be conducting, so that the output is still UP. Only if input 1 *or* input 2 is UP, will the output be up too—hence the name OR gate. If both inputs are DOWN, the output too must be DOWN.

We have shown the basic OR gate with two inputs, but more than two inputs may be used. If we apply the same analysis as in the two-input case, we find that the output will be UP if any input is UP, and DOWN if no input is UP.

The other basic gate is the *diode AND gate* illustrated in Fig. 9.5 (b). Here if either input is DOWN (at 0 V) its corresponding diode is forward biased and the output thus clamped to 0 V is also DOWN. Only if both input 1 *and* input 2 are UP, will both diodes be cut off and the output be UP too—hence the name AND gate. This gate is also known as a 'COINCIDENCE' gate because its output is UP only when UP signals coincide at the inputs.

LOGICAL GATES USING TRIODES

One essential of all digital logic circuits is the '*NOT*' gate illustrated in basic form in Fig. 9.6 (a). This circuit is essentially an overdriven common-emitter amplifier and is often called an INVERTER because it inverts the phase of a signal passed through it (i.e. a positive going signal at the base appears as a negative going signal at the collector). But the NOT gate does more than just phase-invert. It changes the absolute signal level. An UP input gives a DOWN output and vice versa. The basic design feature of the circuit is that $R_B < hfeR_O$ where hfe is a d.c. current gain of the transistor at a collector current V/R_C. Some readers may be puzzled by the description 'gate' attached to this circuit, which does not appear to have any *control* over the input signal, but merely alters its character. This is simply a convention to enable this essential switching circuit block to be classified with other control circuits which are more rigorously described as gates.

With nothing more than the three logic gates so far discussed—diode OR, diode AND, transistor NOT—it is possible to fabricate the complete pulse handling logic of a computer. But this is not the most economical way to do the job. For example, to make up signal losses through the passive diode gates a non-inverting amplifier is required, and for this two inverting NOT gates must be used in series. Transistors, through the diode characteristics of their base-emitter junctions, can be used as active gates to replace the diode OR and AND gates, thus eliminating the need for separate amplifiers.

A basic *transistor OR gate* is illustrated in Fig. 9.6 (b). For the logic levels of 0 and $+V$ selected, n-p-n transistors are used in an arrangement of two emitter-followers with a common emitter load resistor R_O. If input 1 is UP (at $+V$), then transistor Q1 conducts,

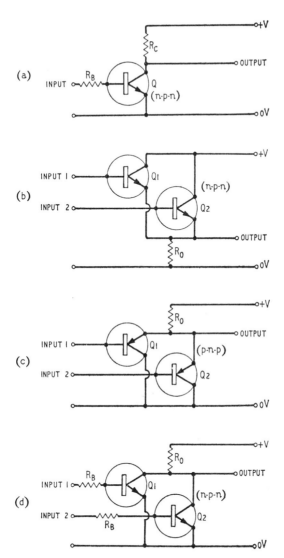

Fig. 9.6. Transistor basic logic gates (for positive level 'up' gates); (a) 'NOT' gate (inverter), (b) 'OR' gate, (c) 'AND' gate, (d) 'NOR' gate

its base-emitter voltage is negligible (ideally zero) and its output at the emitter too is UP. Similarly, if input 2 is UP, the output is UP. Thus if input 1 *or* 2 is UP the output is UP, and we have an OR gate.

The basic *transistor AND gate* takes the form shown in Fig. 9.6 (c). Here, for logic levels O and $+V$, p-n-p transistors are used in an arrangement of two emitter-followers with a common emitter

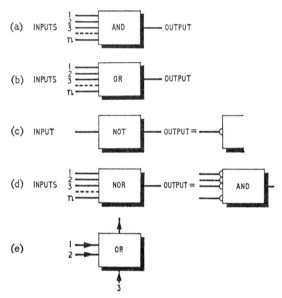

Fig. 9.7. *Symbols for logical gates*

load resistor R_O. If either input 1 or 2 is down (at 0 V), then the related transistor Q1 or Q2 conducts and the output is clamped down to 0 V through the conducting base-emitter diode. Only if *both* inputs are UP (i.e. both input 1 *and* input 2) will the output also be UP. We thus have an AND gate.

One interesting point about these transistor gates is that they use complementary transistor types—n-p-n for the NOT and OR gates and p-n-p for the AND gate. This leads to a logical simplicity that would not be possible with valves (which are equivalent to only one transistor type, the n-p-n). If you restrict your logic gates to only one transistor type, say n-p-n, you can only replicate the more complex valve circuits.

Although we have set out NOT, OR and AND gates as the basic

'Gates' 145

'building block' gates of a digital computing system, in fact these are not all *necessary* gates. There is, however, one gate which by itself is logically complete; that is, any logical expression can be achieved by combinations of this gate. A whole digital handling system could thus be built up with this single gate type—even if it would be uneconomical in practice. This is the '*NOR*' *gate* illustrated in basic form in Fig. 9.6 (d). 'NOR' stands for 'NOT-OR' and implies that the output is *not* UP when either input 1 *or* 2 is UP. This should be clear from the circuit diagram because if either input is UP the positive voltage applied to it biases the corresponding n-p-n transistor hard on and the output voltage at the collector falls virtually to zero (i.e. the output is DOWN). Conversely, when both inputs are DOWN, the transistors are cut-off (because no current flows into either base) and the output is UP. Around a NOR gate of this basic type has been produced a commercial range of logical switching blocks down as 'NORBITS', which can be wired together to produce complete digital switching equipments. In the 'NORBIT' system the basic NOR gate is supplemented by a small number of amplifiers, etc. but the logic is primarily designed round the primary NOR element.

SYMBOLS FOR LOGICAL GATES

Designers of logical switching systems use a system of 'shorthand' rather than draw out the various gates in detail. The gate symbols used have not yet been firmly standardised but one commonly used set is given in Fig. 9.7 where (a) (b) (c) and (d) are the symbols for the AND, OR, NOT and NOR gates so far covered. Where a NOT gate is inserted in an input line to another gate it can be symbolised by the second simplified sign in Fig. 9.7 (c). This can be seen used in Fig. 9.7 (d) where the normal NOR symbol can also be represented as an OR symbol with a NOT symbol in each input line.

The symbols of Fig. 9.7 were selected from the many different types that have been used from time to time, not for any logical 'philosophical' reason but because they are boxes with vertical and horizontal edges which are convenient to outline on a drawing board. Other symbols using circles, or triangles are more troublesome in this respect. The inputs have been shown entering from the left hand edge but they can also enter at top or bottom. Similarly the output can emerge from any side. When it is not self evident which is input and output, appropriate arrowheads are added as in the specimen OR gate at Fig. 9.7 (e).

146 *Elements of Transistor Pulse Circuits*

SUPPLEMENTARY TRANSISTOR LOGIC GATES

A survey of logic gates would not be complete without mention of the two other commonly used special gates shown in Fig. 9.8 together with their symbols—(a) 'INHIBITOR' and (b) 'EXCLUSIVELY OR'.

The INHIBITOR gate of Fig. 9.8 (a) is in effect an AND gate made up of Q1 and Q2 where input 2 is pre-inverted by the NOT gate Q_O. As a result, if input 1 is UP and input 2 DOWN the output will be UP. Conversely if input 2 is UP, then the output cannot be UP whatever the input 1 is. The gate is called an INHIBITOR because an UP signal at the 'inhibiting terminal' 2 prevents an UP signal at input 1 from being transmitted through to the output. The gate is also known as an 'ANTI-COINCIDENCE' circuit, because it gives no output when two inputs coincide. Another self-explanatory name used is 'NOT-AND' or 'NAND' gate. This gate illustrated a point not made clear earlier; namely that any input terminal of a logical gate can be selected for use as a separate control gate and the logical gate then used as a simple transmission gate.

The other auxiliary logic gate illustrated in Fig. 9.8 (b) is the 'EXCLUSIVELY-OR' gate. In the simple OR gate described earlier the output is UP if any one or more of the inputs is UP. Now some logic circuits require the EXCLUSIVELY-OR gate in which the output is UP if one, but not more than one, of the inputs is up. In the arrangement of three gates shown in Fig. 9.8 (b) if inputs 1 and 2 are UP together, the output of the AND gate too is UP and closes the INHIBITOR gate, so that the final output cannot be UP. If either input 1 or input 2 is UP by itself, the output from the AND gate cannot be UP, so that the INHIBITOR gate is not closed. Meantime the output of the OR gate is up because one of its inputs is UP. Thus the final output is UP. Thus this combined EXCLUSIVELY-OR gate gives an UP output when one or other of the inputs (but not both) is UP. This gate is also known as an 'AND-NOT' gate.

PRACTICAL ASPECTS OF LOGICAL GATES

The field of electronics logic has many forms of gate apart from the basic circuits we have looked at above, and only a detailed study of computer circuits can show all the practical problems of their design. We have dealt only with d.c. coupled gates, but in pulse

'Gates' 147

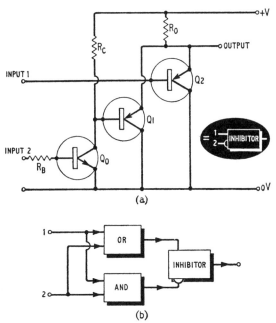

*Fig. 9.8. Supplementary transister logic gates; (a) inhibitor,
(b) exclusively-OR (in symbols)*

circuitry a.c. coupling is frequently used. This brings a need for various d.c.-level-restoring circuits that are outside the scope of this approach survey. Also we have assumed ideal semiconductors with junctions assumed to be short circuits on forward bias and open circuits on reverse bias, thus avoiding such troublesome design problems as leakage current variation with temperature. Readers interested in more exact design-information should look up such a treatise as J. Millman and H. Taub *Pulse and Digital Circuits* (McGraw Hill 1952).

While we have looked briefly at diode and transistor gates, it should not be overlooked that in recent years a new series of devices such as tunnel diodes, p-n-p-n 'thyristors', s.c.r.s. etc. are opening up new types of transmission and logic gates to supplement the basic diode and transistor triode types. As yet these are relatively new and no easily accessible complete accounts of their use as gates are available, but it is probable that they will change the face of gating practice as much as the transistor did when it 'took over' from the valve.

CHAPTER 10

Counter/Timers (Frequency Meters)

Nowadays, if you work in an electronics laboratory, you are expected to have some knowledge of counter-type frequency measuring instruments. These are steadily moving into the status of essential laboratory equipment—alongside the multi-meter, the signal generator and the oscilloscope. Many basic pulse circuits discussed previously find a place in such counters, and looking at the working of a counter conveniently shows them in action.

Until the invention of the electronic counter, precise frequency measurements were carried out by comparing the unknown frequency with some accurately known variable standard, zero-beating the standard against the unknown. Heterodyne frequency measurements of this type tended to be slow, difficult, ambiguous and expensive. The electronic counter has changed this. It has placed within the reach of the average small laboratory equipment which is capable of measuring frequencies accurately to say seven or eight significant figures (e.g. to 1 cycle in 10 MHz), and yet of giving these precise results almost instantaneously, without ambiguity and with almost no manipulation of apparatus.

The basic principle of the electronic counter-timer is simply to count the number of cycles of an unknown frequency that occur in unit time. Fig. 10.1 illustrates this in block diagram form. To measure a frequency, the signal is fed into a gate as shown in Fig. 10.1 (a). This gate is controlled by a timebase in the instrument, which arranges to open the gate for a precisely determined length of time, say 1 second. During the second that the gate is open, the input signal passes through and records on the counter the number of cycles passed in that second. The gate then closes and the number of cycles is left recorded and visually displayed on the counter readout. Thus we can read directly on the counter the frequency of the input signal with the potential accuracy of the number of digits displayed. For example, with six-digit displays it is possible to count up to 999,999.

The same equipment can be used also to measure time or period, rather than number or frequency. To do this, it is arranged as in Fig. 10.1 (b). Here the signal input is used to control the opening

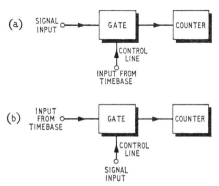

Fig. 10.1. Basic principle of counter/timer. (a) Measuring number or frequency, (b) Measuring time or period

and shutting of the gate, and the input from the timebase is fed through to the counter. The signal input opens the gate for the duration of one period of the signal frequency. If the timebase is operating at say 1 MHz frequency, it will be supplying one pulse every micro-second into the gate. Thus there will be recorded on the counter the number of microsecond pulses passing through in one cycle of the signal input; the counter will then read in microseconds the period of the signal frequency.

We have illustrated the principle of the counter/timer in terms of a periodic signal input. However, it can be used equally well to measure the total number of impulses coming in on the signal input line (Fig. 10.1 (a)), even though they do not recur at a regular periodic rate. Similarly in the case of Fig. 10.1 (b) the equipment can be used to measure the total time between two isolated events which are fed into the signal line, one opening the gate and the other closing it.

TYPICAL ELECTRONIC COUNTER/TIMER

Counter-timers (or digital frequency meters as they are sometimes called) in their commercial versions show many detailed differences of circuitry, but stripped of unessentials they all reduce in essence to the arrangement shown in block diagram form in Fig. 10.2. After the frequency to be measured has been fed into the input, the *input pulse shaper* amplifies the signal and converts the waveshape into the standard form of a square wave with very fast rise and fall times, suitable for driving an electronic counter. The

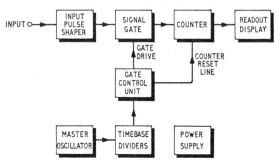

Fig. 10.2. Simplified diagram of typical electronic counter/timer

shaped pulses are permitted to pass through the *signal gate* for such time as it is opened by the *gate control unit*. The whole system has been reset to zero before the beginning of the gate-open period so that the number of pulses passing through the signal gate are stored in total in the *counter*. The counter contents are recorded in visual numerical form in the *readout display*.

Thus far the gate control unit has performed two functions. It has reset the counter to zero, and opened the gate for the required gate period. The control of the gate period is effected from a timebase. This timebase comprises (a) *a master oscillator* which is a stable crystal-controlled circuit and (b) a set of decade counter 'times-ten' *dividers*, usually referred to as DCUs. The master oscillator itself is sinusoidal, but it includes a buffer squaring circuit which provides a timebase square wave of frequency f. The timebase DCUs provide derived square-wave outputs at frequencies $f/10$, $f/100$, $f/1,000$ etc. The operator selects the desired timebase frequency from those available and switches this manually into the gate control unit.

The *power supply* shown separately in Fig. 10.2 is the unit which supplies all the d.c. rail voltages to the other sections of the counter-timer. As it has no particular features of interest in the pulse circuitry aspect, we will not consider its detailed design. It is worth noting, however, that with the modern trend to transportable instruments for use independently of mains supplies, commercial units nowadays show a tendency to end up with a 12 V supply which can be interchangeably provided by a car battery (for mobile work) or derived from the mains (for fixed station operation).

In the discussions of the previous paragraphs, we have considered that the gate has opened for only a single period of the timebase and closed again. However, if the unknown frequency varies with time, a single sample of this sort will not be satisfactory. The gate control unit has therefore to provide for successive repetitive sampling of the input signal so as to correct continuously the readout display of the frequency. This will be discussed in greater detail later.

COUNTER

The most important and probably the least understood section of a frequency meter is the *counter* itself. This usually comprises a counting chain of decade counter units arranged in cascade so that each one divides the output of the previous one by ten. Each DCU is a cascade of four binary counters. You can follow the build-up from the basic binaries to the complete counter in Fig. 10.3.

A typical basic *binary* element is illustrated in Fig. 10.3 (a). This type of circuit has been discussed in some detail in previous chapters in this series. It is a bistable multivibrator of the Eccles-Jordan type. Pulses fed into the input line are steered alternately to the two transistor bases in turn and cause the circuit to switch from the condition where the right-hand transistor is conducting to cut-off and back again. We take the condition with the right-hand transistor conducting as the 'set' state of the binary and where it is cut off as the 'unset' state. Every time the binary is driven to the set state a positive-going pulse appears at the output, and to the unset state a negative-going one. The reset line shown is normally shorted to the earth line, but, if it is disconnected from earth, the right-hand transistor is driven hard on through the 470 ohms and 6·8 kΩ connected from the negative rail to its base. Thus to reset the binary, all that is necessary is to open circuit the connection between the reset line and earth. The actual example of a binary given in Fig. 10.3 (a) is capable of switching reliably at not less than 1 MHz.

152 *Elements of Transistor Pulse Circuits*

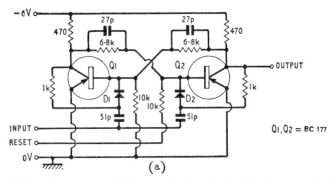

Fig. 10.3. Counter unit build-up. (a) *Typical 1 Mc/s binary.* (b) *Four binaries connected as decade counter unit DCU.* (c) *Six DCU's cascaded as six-digit counter*

The *decade counter unit* shown in diagrammatic form in Fig. 10.3 (b) is an arrangement of four binaries with two feedback loops. The output of each binary feeds into the input of the next one. Each binary divides by two, so that the cascade of four would without feedback divide by sixteen, i.e., would give one output pulse for each sixteen that are fed into the input. To divide by ten, two feedback loops are inserted which cause the decade counter unit to skip altogether six counts. For every ten input pulses then, the DCU gives one output pulse. There are many ways of arranging feedback loops to convert a divide-by-sixteen into a divide-by-ten. The method of Fig. 10.3 (b) is only one of the many possibilities. The output terminals marked '$\bar{1}$', '1', '$\bar{2}$', '2', etc., in Fig. 10.3 (b) are taken from the collectors of the respective binaries. They are used to drive a separate *readout* display which shows numerically by some form of lamp or illuminated figure the actual number of counts stored in the DCU. These may vary from 0 to 9. When the contents of the DCU reach 10, it gives out a pulse and reads 0 again to carry on counting the next ten pulses.

The complete *counter* unit is made up of a cascaded series of DCUs as shown in Fig. 10.3 (c), each one feeding its output into the input of the next one. The same number of DCUs are required as the number of digits required in the readout.

READOUT DISPLAY

The *readout display* shown in the layout of Fig. 10.2 as fed from the counter can take many forms in practice, varying from a cheap

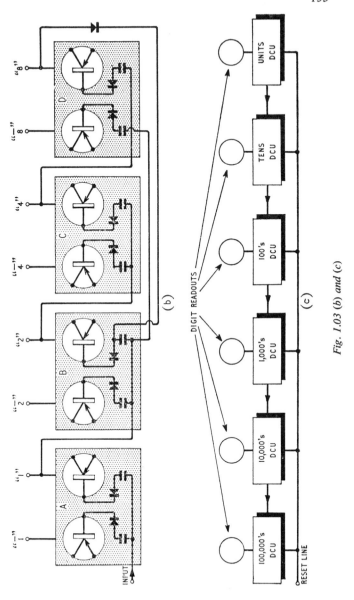

Fig. 1.03 (b) and (c)

154 *Elements of Transistor Pulse Circuits*

meter for each digit place, up to very complex electroluminescent elements. Much ingenuity has been consumed in trying to produce a cheap reliable readout device, but with no great success so far. Most of the commercial readout units are expensive in themselves and also require an expensive decoding network to transform the binary information from the counters into a decimal display. Because there is little standardisation in this field yet, we will not consider further here the decoder circuits, etc., required in a practical instrument.

INPUT PULSE SHAPER

Logically the next element to be considered in the layout of Fig. 10.2 is the *input pulse shaper*. Fig. 10.4 gives a detailed circuit of a typical pulse shaper capable of operation up to 1 MHz and over. Here the input signal passes through a 4·7 kΩ resistor attenuator with a 10 pF compensating shunt capacitor for operation up to the highest frequency handled. At point A the signal is rectified by the diode D1 and the base-emitter diode of transistor Q1. After amplification through Q1, the signal reappears rectified at the collector as shown., i.e., in the form of a semi-sinusoidal pulse for each cycle of the input. Applied to the input of the Schmitt trigger Q2, Q3, the signal finally appears at the output of Q3 as a steep-sided squarewave pulse suitable for driving the counter circuits.

Bias to the base of the first transistor Q1 is provided from the 50 kΩ potentiometer across the $-$ 6 V, $+$ 3 V lines. By varying the setting of the slider on this potentiometer, it is possible to set the bias point of Q1 from full cut-off to full conduction. By this means we can adjust the setting of Q1 for optimum trigger sensitivity for any amplitude of input signal. In most commercial equipments the pulse shaper circuit is like the typical example given, i.e., a rectifying amplifier followed by a squaring Schmitt trigger.

SIGNAL GATE

The *signal gate* shown in block form in Fig. 10.2 as controlling the transfer of pulses from the input pulse shaper to the counter can take many forms, depending on the type of gate circuit favoured by the designer. Fig. 10.5 shows a typical circuit for this position. The output from the pulse shaper is fed via an 8·2 kΩ isolating resistor into the base of the transistor Q1. As before, the input signal is rectified by the diode D1 and the base-emitter diode of Q1.

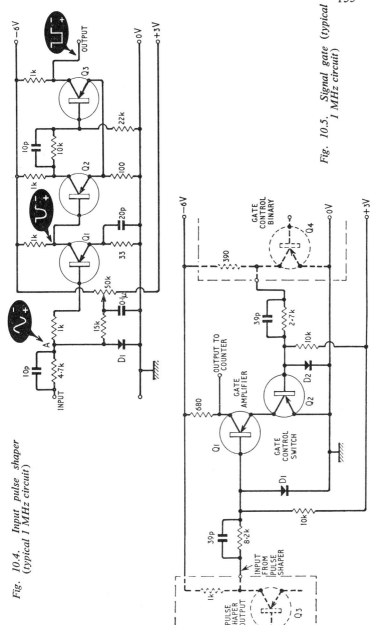

Fig. 10.4. *Input pulse shaper (typical 1 MHz circuit)*

Fig. 10.5. *Signal gate (typical 1 MHz circuit)*

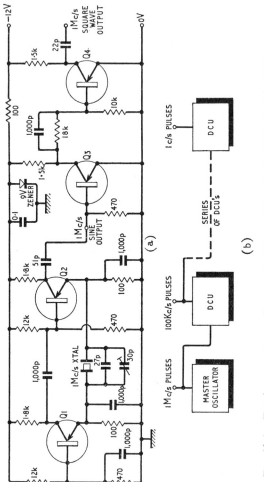

Fig. 10.6. *Timebase master oscillator and dividers; (a) typical 1 MHz high stability master oscillator, (b) DCU's used to provide range of timebase frequencies by repeated tens-division from 1 MHz*

Counter/Timers (Frequency Meters)

This rectified input waveform provides an amplified output unidirectional pulse at the collector of Q1. This gate amplifier can operate as a common emitter amplifier and thus pass a pulse through only if its emitter is effectively short-circuited to deck by the control gate switch transistor Q2 being bottomed (full on with a relatively low resistance from its collector to emitter). When the base of Q2 is negative the transistor is driven hard on and the gate amplifier can operate, i.e., the gate is open. Conversely, when the base of Q2 is positive, the transistor is cut off, and its collector presents a high impedance to the collector of Q1. This means that the gate is then closed.

The control voltage for the base of Q2 can be provided from the collector of the gate-control binary Q4, through a dropping resistor of 2·7 kΩ and a 10 kΩ resistor to the positive rail. When the gate control binary Q4 is cut off its collector rises close to -6 V and the base of Q2 connected to the centre point of 2·7 kΩ, 10 kΩ attenuator from -6 V to $+3$ V is at a negative potential with respect to deck. Q2 is then switched hard on and the gate is open. Conversely it can be shown that, when the gate control binary Q4 is switched hard on, the collector is virtually at deck potential, and the base of Q2 is therefore positive. This cuts off Q2 and closes the gate.

MASTER CONTROL OSCILLATOR

Considering the elements in Fig. 10.2 further, the next one to look at is the *master oscillator* unit. As mentioned earlier this is usually a crystal-controlled oscillator feeding into a squaring circuit unit. A typical 1 MHz master oscillator circuit is given in Fig. 10.6 (a). Q1 and Q2 form a Butler crystal oscillator circuit with positive feedback from the emitter of Q2 to the emitter of Q1 via the tuned circuit comprising the 1 MHz crystal shunted by the fixed capacitance of 27 pF and the variable 30 pF. The variable capacitor is used for fine adjustment of the oscillator frequency. The 1000 pF capacitors connected from the emitters of Q1 and Q2 to deck may look like decoupling capacitors across the 100 Ω emitter resistors which would prevent the circuit oscillating. However, the decoupling is not complete because at 1 MHz 1000 pF has 150 Ω impedance and the resultant impedances in the emitter circuits are not less than 60 Ω each. The sinusoidal output of Q2 is taken via a 51 pF into the base of the overdriven amplifier Q3 followed by a further d.c. coupled amplifier Q4, which between them square off the

158   *Elements of Transistor Pulse Circuits*

sine wave input and give a 1 MHz square wave output from the collector of Q 4. To achieve high oscillator stability, the crystal is normally mounted in a constant-temperature oven. For stabilisation against supply voltage changes, the d.c. supply to the two oscillator transistors is provided from a 9 V Zener-stabilised rail as shown in the diagram.

TIMEBASE DIVIDERS

The master oscillator by itself can provide only a 1 μsec gate time, and to deal with lower frequency input signals outputs at 100 kHz, 10 kHz, etc., down to 1 Hz are normally required to provide gate times of 10 μsec, 100 μsec, etc., up to 1 second. The *time-base dividers* section of Fig. 10.2 is usually arranged as shown in Fig. 10.6 (b). Here the master oscillator square-wave output at 1 MHz is fed into the first DCU, which divides by ten and gives a 100 MHz square-wave output. This again is fed into a further string of DCUs, each of which in turn divides by ten. The outputs from these dividers provide trains of square waves at decade intervals down to 1 Hz. Any of these can be selected for controlling the open time of the signal gate.

CONTROL UNIT

We now come in Fig. 10.2 to the logical heart of the frequency meter, the gate *control unit*. This is the most complex part of the equipment and in Fig. 10.7 is set out in diagrammatic form the control unit of a typical counter/timer to use as an illustration of how it works. Designs may differ in detail from Fig. 10.7 but all follow pretty much this pattern. The control unit is designed primarily to control the signal gate shown dotted at the top of the diagram, and its operations are controlled partly by the output from the timebase shown coming in on the left of the diagram. To illustrate the operation, let us assume to start with that the signal gate is closed. Let us also assume that the various binaries have all been reset in the control unit and the counter.

The first pulse from the timebase arriving at the open timebase gate passes through to the gate control binary and flips it over. The flipped binary feeds a d.c. signal through the signal buffer amplifier to open the signal gate. This lets the stream of signal pulses start feeding into the counter. The next timebase pulse arriving still finds the timebase gate open and flips the gate control

Counter/Timers (Frequency Meters) 159

binary back to its original state. This cuts off the d.c. control voltage to the signal gate and closes it. The stream of signal pulses to the counter is cut off, and the counter holds and displays the number of pulses that it has received while the gate was open.

In flipping back to its original 'set' state, the gate control binary sends a positive pulse to the latch binary which causes it in turn to

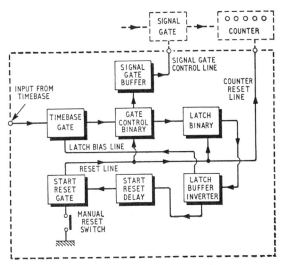

Fig. 10.7. Control unit of counter/timer

flip over. In flipping over, the latch binary sends a negative d.c. voltage to the latch buffer amplifier inverter. This voltage is inverted in the latch buffer and appears as a positive voltage to close the timebase gate so that further timebase pulses are inhibited (prevented passing through into the system) until the gate can be reopened.

If we want the counter/timer to sample the input signal repetitively we must arrange for the whole system to be automatically reset to the initial state to let the cycle repeat again and again. This is achieved by using the same positive d.c. voltage shift from the latch buffer inverter that was used to close the timebase gate also to trigger off the start reset delay unit in Fit. 10.7. This circuit is a pulse delay unit which emits an output pulse at a predetermined interval after it receives the input pulse. The delayed output pulse from the start reset delay is used to open-circuit the start reset gate in the reset line. The opening of the start reset gate removes the

160 *Elements of Transistor Pulse Circuits*

earth connection of the reset line and resets the latch binary and the DCUs in the counter. In being reset, the latch binary emits a negative d.c. voltage level change which (inverted through the latch buffer inverter to a positive d.c. voltage level change) opens the timebase gate once again and lets the counting cycle repeat 'from scratch'. As the start reset delay circuit is arranged to respond only to a positive going pulse, it is not affected by this output from the latch buffer inverter.

Besides the automatic start reset provided by the start reset delay circuit, equipments usually include some form of manual reset

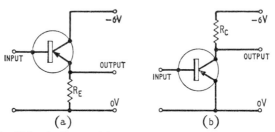

Fig. 10.8. Buffer amplifiers; (a) emitter follower, (b) inverter

switch as shown in Fig. 10.7. This is normally closed, but can be open-circuited manually to reset the whole counter ready to start counting.

The complete operation of the control unit may seem rather complex at first sight, but careful following through of the logic will soon show how the various circuit blocks work together.

BUFFER AMPLIFIER AND INVERTER

Before we go on to consider the start reset delay circuit in detail, we will first take a look at how the two *buffer amplifiers* in Fig. 10.7 work.

The *signal gate buffer* amplifier is an emitter follower of the type illustrated in basic form in Fig. 10.8 (a). The output is in phase with the input, so that the negative output voltage from the gate control binary is transmitted to the signal gate also as a negative voltage, suitable for opening the gate.

The *latch buffer inverter* in Fig. 10.7, on the other hand, is a common emitter amplifier of the type shown in Fig. 10.8 (b) which inverts the phase of the input signal and provides from the negative input voltage fed to it from the latch binary the necessary output positive voltage to trigger on the start reset delay circuit.

161

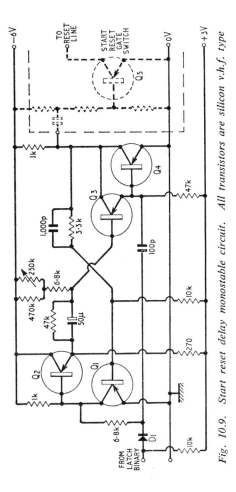

Fig. 10.9. Start reset delay monostable circuit. All transistors are silicon v.h.f. type

START RESET DELAY

This brings us finally to detailed consideration of the last remaining circuit in Fig. 10.7, i.e., the *start reset delay* circuit. This is usually some form of monostable, and generally it is a multivibrator. Fig. 10.9 gives the circuit of a typical example to illustrate the design points. Q1 and Q4 together form a monostable multivibrator, with resistive cross-coupling from collector to base in one direction and capacitor cross-coupling in the other. The collector of Q4 is cross-coupled to the base of Q1 via the 3·3 kΩ resistor. The 1,000 pF capacitor across the 3·3 kΩ is merely a speed-up capacitor, negligible compared with the resistor. The capacitor cross-coupling from the collector of Q1 to the opposite base is via the emitter follower isolating transistor Q2 through the 50 μF coupling capacitor to the base of Q3. Q3 and Q4 form a Darlington pair equivalent to a single very high gain transistor.

The circuit operates roughly as follows. In the quiescent condition, the transistor Q4 is switched hard on by the 6·8 kΩ, 470 kΩ and variable 250 kΩ network to the negative supply rail. When a positive pulse comes in from the latch binary, it passes through the diode D1 and the 100 pF capacitor to the base of Q4. This switches transistor Q4 off and Q1 on. A negative-going pulse appears at the collector of Q4 and is transmitted through the output coupling capacitor to the base of the start reset gate switch transistor Q5. Now Q5 is normally held switched hard on by the bias network of resistors to its base, and the additional negative pulse merely drives it harder on. Q5 thus remains fully bottomed, keeping the reset line shorted to deck.

During the quasi-stable state of the monostable, the 50 μF capacitor discharges steadily with a time-constant set by the variable network of resistors connected to its right-hand end. Eventually, at a time determined by the setting of the 250 kΩ variable resistor, the monostable flips back to its stable condition. When this happens, a positive pulse appears at the collector of Q4, is transmitted to the base of Q5 and cuts that transistor off for a short time. This open-circuits the reset line from deck and causes all the binaries in the control unit and the DCUs in the counter to be reset.

When the latch binary is reset in Fig. 10.7 the negative voltage on its output disappears and similarly the positive inverted voltage at the output of the latch buffer inverter which has been holding the timebase gate closed also disappears. Thus the timebase gate is reopened.

The whole system is thus reset ready to start as before, and the next input pulse from the timebase triggers off the counting cycle once again. The repetition time between sample counts is set by the 250 kΩ variable resistor control on the start reset delay unit. In a practical instrument the recycling time is normally variable between a small fraction of a second and some tens of seconds.

CONCLUSION

In this chapter we have attempted to set out the main operational features of a typical counter-type frequency meter as a demonstration of uses of some of the basic elementary pulse circuits dealt with in earlier chapters. In the space available it has not been possible to go into all the possible circuit variations likely to be met with in a commercial counter/timer, but most counter/timers work very much along the main lines indicated above.

CHAPTER 11

Timebases (Sweep-Generators)

'Miller integrators', 'bootstrap circuits', 'phantastrons'—as soon as the budding engineer starts to take an interest in time bases he runs into a discouraging array of jargon used by the *cognoscenti*. He oftentimes gives up despairingly in what is not really a very difficult area of pulse circuits. This chapter is an attempt to lead him gently to some understanding of the 'arcana'.

If you turn to O. S. Puckle's classic *Timebases* (Chapman and Hall Ltd.) you will find a fascinating detailed account of many different timebases likely to be met with in valve practice. With transistors, we will simplify matters by confining our attention to what are called 'linear timebases', i.e. circuits producing a waveform that rises steadily with time until it reaches a certain amplitude and returns in a much shorter time to the point from which the linear rise can start again. The forward linear rise is usually called the 'sweep' and the sharper return the 'flyback'.

Different authors use different names to describe this type of generator, e.g. 'timebase', 'sweep generator', 'ramp generator', and 'sawtooth generator'. Now a timebase output can be either a current or a voltage. Again it can be single-sided or symmetrical push-pull; i.e. it can be a unidirectional sawtooth or can be made up of two sawtooths of equal and opposite polarity. The second type is often required for balanced driving of c.r.t. deflection circuits.

Timebases find their principal use in sweeping the beam of an oscilloscope across the face of the cathode ray tube to make possible the display of a time-varying signal. In this context, we find them widely used in radar and television circuits, computers, time-measuring devices, and time-modulation equipment. They also find a place in analogue and digital equipment, e.g. in some digital voltmeters.

DRIVEN NON-REGENERATIVE TIMEBASES

The simplest type of linear timebase is the circuit which takes in a square wave and gives out a sawtooth, the forward sweep of the output occupying the same time as the input pulse. When the input pulse ceases, the flyback occurs and the circuit returns to its quiescent condition. This sort of circuit is commonly non-regenerative. It really just reshapes the input square wave into a ramp voltage. A whole family of elementary circuits can be used as the basis of this type of non-regenerative timebase.

'Transistor switch' timebase

The most primitive form of non-regenerative linear timebase uses a transistor as a simple switch to control an RC relaxation network. The basic circuit arrangement is shown in Fig. 11.1 (a). Here the transistor Q is set up so that, in the absence of any signal at the input, it is bottomed. This means that the current through the base resistor, R_B, holds the transistor fully switched on, and the collector

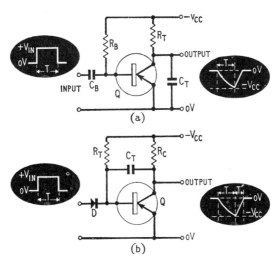

Fig. 11.1. Simple transistor sweep generator (driven by square-wave input); (a) without feedback, (b) with feedback

166 *Elements of Transistor Pulse Circuits*

output potential is close to zero volts. In this condition, the capacitor C_T is discharged. If now a positive pulse of duration T is applied to the input *via* the large coupling capacitor C_B, the transistor base is immediately driven positive and the transistor cuts off for the duration of the pulse. With Q cut off, the capacitor C_T begins to charge negatively through resistor R_T. This is shown in the output waveform in Fig. 11.1 (a) for the interval T. At the end of the pulse, the positive voltage is removed from the base, and the transistor is driven on again through R_B. The transistor bottoms and presents a low impedance across the capacitor C_T, causing it to discharge rapidly, and bringing the collector potential quickly back close to zero volts again as shown. The output sweep voltage during the interval T is exponential in form, but, provided the pulse time T is small compared with the time constant $C_T R_T$, the sweep will be a reasonable approximation to a linear waveform.

Another circuit which on paper looks very similar to the last one is given in Fig. 11.1 (b). This too is driven by a positive square-wave input pulse, but the action is somewhat different due to the negative feedback capacitor C_T, connected between the transistor collector and base. As before, in the quiescent condition, the transistor is held hard on by the base current through the resistor R_T from the negative supply rail. When the pulse arrives, it takes the base positive and cuts the transistor off. The capacitor C_T then begins to charge up through R_C with a time constant $C_T R_C$ as shown for the interval T in the output waveform of Fig. 11.1 (b). When the input pulse ends, the positive voltage is removed from the base, and the transistor base drive current through R_T is restored. The transistor does not, however, return instantaneously to the bottomed condition where its potential is close to zero volts, because its collector potential is held up initially by the charge on C_T. C_T then discharges through R_T with a time constant proportional to $C_T R_T$, as shown in the output waveform for the second interval marked T'. This linear 'secondary' waveform can be used as the forward sweep voltage of a timebase, and its rise time is a function of C_T, R_T and the h_{FE} of the transistor.

This circuit is a sort of 'phantastron'. In valve practice the 'phantastron' was a circuit in which the time base was triggered off by a short pulse at the input, and then the circuit supplied its own gating to continue the linear sweep. The pulse merely initiated the sweep, which then 'runs down' without any further help from the trigger input. The circuit of Fig. 11.1 (b) really belongs more correctly to the family of Miller integrators to be described below.

Miller integrator timebase

The circuit of Fig. 11.2 (a) is the Miller integrator sweep generator. This widely-used circuit works as follows. It is biased so that when the switch Sw is closed (as it normally is) the transistor Q is held cut off, because the base is short-circuited to earth, and no base drive current flows into the transistor. Thus the collector is virtually at the negative rail voltage and the base at 0 V. As a result, C_T is negatively charged to $-V_{CC}$. If now the switch is opened at time

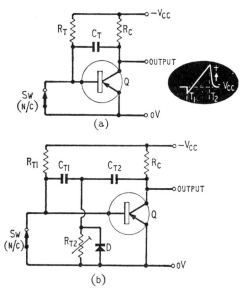

Fig. 11.2. Miller sweep generator; (a) simple basic circuit, (b) with additional linearing network

T_1 and closed at time T_2, the voltage at the collector of the transistor takes the waveform shown. Immediately the switch is opened, the base goes a few hundred millivolts negative, when the input current is switched into the base through R_T. This small negative step voltage is transmitted *via* C_T to the collector, and appears as a small negative-going jump at the point T_1 in the output waveform (which is characteristic of the basic Miller integrator). This can be shown to be due to the finite output resistance of the transistor, and can be reduced to zero by including a suitable resistor in series with C_T. In the simple version shown, once Sw has been opened, C_T begins to discharge through R_T, with a time constant approximately

168 *Elements of Transistor Pulse Circuits*

proportional to $C_T R_T$. However, the ramp output waveform corresponding to this is not exponential but linear. This is because of the feedback action explained in an earlier chapter on operational amplifiers. During the process of discharge, the current through C_T remains constant. When the switch Sw is closed again at time T_2, the transistor is switched off again and C_T discharges through R_C, to give a waveshaping returning finally to $-V_{CC}$ as shown. R_C is usually much smaller than R_T so that the recovery or flyback time is much shorter than the forward sweep time.

Although the Miller integrator circuit produces a ramp waveform much more linear than that which would be achieved merely by discharging a capacitor through a resistor, its linearity can be even further improved by additional feedback circuits. A commonly used one is illustrated in Fig. 11.2 (b). Here the feedback timing capacitor C_T has been split into two capacitors C_{T_1} and C_{T_2}, and a resistance R_{T_2} taken from their midpoint to earth. Very commonly C_{T_1} and C_{T_2} are made equal and R_{T_2} is made variable so that the linearity can be adjusted. The diode D across R_{T_2} serves the purpose of speeding up the recovery time of the circuit.

In both versions of the Miller circuit given, the switch Sw is shown as a normally closed mechanical switch. In practice a transistor is very often used for this purpose. In the quiescent condition this switch transistor is held bottomed, and its collector effectually presents a short circuit to the base of the Miller circuit transistor. When the switch transistor is switched off, it becomes effectually an open circuit, and we have a direct equivalent of a mechanical on-off switch.

Bootstrap sweep generator

Another very common non-regenerative driven sweep generator circuit is the 'bootstrap' integrator. In its simplest form this takes the shape shown in Fig. 11.3 (a). It is essentially an emitter-follower with feedback from the emitter *via* a capacitor C_F to the junction point of two resistors R_F and R_T feeding current into the transistor base, with a capacitor C_T from the base to earth. The capacitor C_T is normally short-circuited by the switch Sw. The transistor base being thus connected to earth, the transistor is cut off, and its emitter is at zero volts. If the switch Sw is now opened at time T_1, the output waveform takes the shape shown in the diagram. C_T begins to charge up through R_F and R_T. As the voltage on the base of the transistor begins to go negative, so does the emitter output voltage

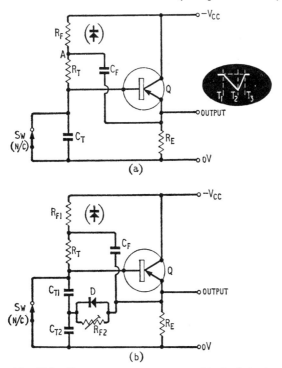

Fig. 11.3. Bootstrap sweep generator; (a) simple basic circuit, (b) with additional linearising network

through emitter-follower action. This emitter voltage change is transferred through the large feedback capacitor C_F to the top end of the resistor R_T, point A. As a result the voltage across the resistor R_T remains virtually constant, and provides a fixed charging current for the timing capacitor C_T. The output voltage follows the base voltage closely and we have the linear ramp shown in Fig. 11.3 (a). At time T_2 the switch Sw is closed again. The transistor then cuts off, and the capacitor C_T discharges rapidly. This gives the flyback path from T_2 to T_3 shown in the circuit diagram illustration. The resistor R_F is often replaced by a diode as shown in brackets to remove the limitation on the recovery time which is set by the charging time constant $C_F R_F$.

The linearity of the sweep voltage is good in the bootstrap circuit, but it can be improved even further by using some form of compensated feedback circuit. Typical of such linearising circuits is the

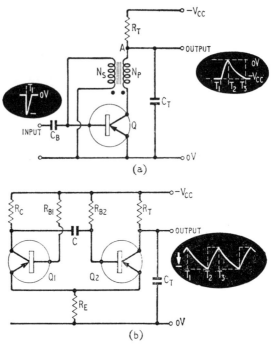

Fig. 11.4. Regenerative sweep generators; (a) *blocking oscillator,* (b) *multivibrator*

arrangement shown in Fig. 11.3 (b). Here the timing capacitor C_T is split into two parts C_{T_1} and C_{T_2}, and feedback is applied to the centre point via a resistor R_{F_2}. The similarity to the circuit of Fig. 11.2 (b) is immediately evident. The shunting diode D is again to speed up recovery time. A common rule of thumb is to make C_{T_1} and C_{T_2} equal, and use a variable R_{F_2} to adjust for best linearity.

BASIC TRIGGER SWEEP GENERATORS

Up till now we have been considering mainly non-regenerative circuits, where the output sweep continues only as long as the input square wave continues. There is also, however, a family of regenerative timebases where the action is merely triggered off by a short pulse and continues thereafter on its own. The commonest of these are the blocking oscillator and the multivibrator.

Timebases (Sweep-Generators) 171

Blocking oscillator sweep generators

In an earlier chapter we discussed in some detail the operation of the blocking oscillator circuit. One use of this circuit as a sweep generator is demonstrated in basic form in Fig. 11.4 (a). Here the transistor base is normally d.c. short-circuited to earth *via* the transformer feedback winding, so that the transistor is cut off. The transistor collector and the output point A are both virtually at negative rail potential. Capacitor C_T is therefore charged up to $-V_{CC}$. If now a negative trigger pulse is applied at time T_1 through capacitor C_B to the transistor base, blocking action sets in and the transistor bottoms firmly for the blocking period of the oscillator. The capacitor C_T is rapidly discharged and the output voltage at point A moves rapidly to near zero volts, remaining there until time T_2, the end of the blocking pulse. The transistor now cuts itself off and the capacitor C_T begins to charge up negatively through R_T towards the rail voltage, as shown in the second part of the waveshape in Fig. 11.4 (a), eventually returning to $-V_{CC}$ at time T_3. During the interval T_2, T_3 the output voltage provides a ramp suitable for sweep generator use. The sweep is, of course, exponentially non-linear because it merely represents the charging up of the capacitor C_T through a resistor R_T from a fixed voltage $-V_{CC}$. Various circuits can be used, however, to linearise the sweep output.

Multivibrator regenerative sweep generators

By the same sort of arrangement where a timing capacitor is shunted across an output resistor, the multivibrator (also discussed in previous chapters) can be used for sweep generation. One example of this is given in Fig. 11.4 (b). Here an emitter-coupled astable multivibrator is set up with a timing capacitor C_T connected from the right-hand collector to earth. Without the capacitor C_T, the output would have the waveshape shown dotted in the diagram. The addition of the capacitor changes the output to the sawtooth waveshape shown. From T_1 to T_2 the transistor Q_2 is cut off, and C_T charges negatively through R_T with an approximate time constant $C_T R_T$. From T_2 to T_3 when the transistor Q_2 is switched hard on, the capacitor C_T discharges through R_E. As R_E is usually designed to be much lower in value than R_T, the recovery time T_3-T_2 is much shorter than the forward sweep time T_2-T_1. By suitable choice of the time constants in the multivibrator and C_T it is possible to make the recovery time much shorter than the sweep time, and give a fair approximation to a linear sweep generator. Of course, without feedback, the sweep must be quasi-exponential.

SWEEP GENERATOR LINEARITY CONSIDERATIONS

We have skated rather lightly over linearity considerations up to this point, but as they are of great importance in practical applications, we must take a closer look at the requirements in this respect. In the case of a general purpose c.r.t. sweep circuit, one important requirement of the sweep is that the sweep speed, i.e. the rate of

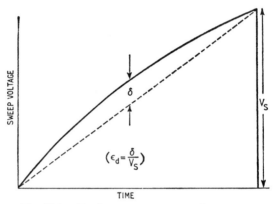

Fig. 11.5. Displacement errors, ϵ_d in linear sweep

change of output signal with time, be constant. A useful definition of deviation from linearity in this case is the ratio of the difference in slope at the beginning and the end of the sweep to the initial value. This is known as the slope (or sweep-speed) error and is usually denoted by the symbol ϵ_s.

In connection with other timing applications, however, a more important index of linearity is the maximum difference between the actual sweep signal and the linear sweep which passes through the beginning and the end points of the actual sweep as shown in Fig. 11.5. This index is called the displacement error, usually denoted ϵ_d, and defined as the ratio of the maximum deviation of the actual from the linear sweep to the final sweep amplitude. It is usually expressed as a percentage rather than as a fraction.

Another non-linearity index sometimes met with is the transmission error, ϵ_t. This arises when a sweep voltage is transmitted through an RC high-pass network and the output falls away from the linear sweep towards the end. The transmission error is defined as the ratio of the output 'droop' to the input signal measured at the end of the sweep.

Provided the deviation from linearity due to the phenomena mentioned is small, so that the actual sweep voltage may be approximated by the sum of a linear and a quadratic term in T, it can be shown from the above definitions that to a first approximation

$$\epsilon_d = \frac{\epsilon_s}{8} = \frac{\epsilon_T}{4}$$

In the remainder of this discussion, therefore, we will confine our considerations to the displacement error, ϵ_d, to simplify matters.

To put some values to the errors discussed above, so you can see what sort of linearity to expect, it can be taken as a rule of thumb that for a general-purpose oscilloscope a displacement error of about 1·0–1·25% is tolerable. Now for the straight exponential charging-up of a capacitor C through a resistor R from voltage V_S, it can be shown that the displacement error is $V_{max}/(8\ V_S)$ where V_{max} is the voltage to which the capacitor is allowed to charge by the end of the sweep. If the sweep time is substantially less than the CR time constant, the displacement error can then be shown to be approximately $T/(8RC)$. Hence, if the sweep is to be reasonably linear the time constant RC must be large compared with the sweep time, T. In general-purpose oscilloscopes, the circuits are usually arranged so that V_{max} is about a tenth of V_S. This gives a displacement error of about 1·25% (as noted earlier).

In many timing applications, however, a much higher precision than this is required, and it is usually then necessary to go to the Miller integrator or bootstrap circuits. As a rough approximation, it can be taken that with a Miller integrator the linearity is improved by a factor of A, where A is the voltage gain of the Miller amplifier. Similarly in the case of the bootstrap, the linearity is improved by a factor of $1/(1-A')$, where A' is the voltage gain of the emitter follower. As A' is slightly less than unity, $1/(1-A')$ is quite large and will be of the same order of magnitude as A in the Miller integrator case. This illustrates clearly why in the main Miller and bootstrap integrators are used in high-precision sweep generators.

Apart from the basic circuit arrangement, various compensating networks are added in practice to improve linearity even further. Two of these have been illustrated earlier in Fig. 11.2 (b) and 11.3 (b).

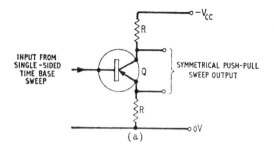

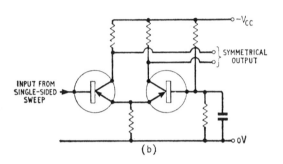

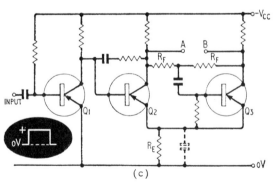

Fig. 11.6. Symmetrical sweep generators; (a) phase splitter inversion of single-ended sweep, (b) long-tailed pair balanced inversion, (c) separate phase-inversion, transistor

SYMMETRICAL TIMEBASES

So far we have been concerned mainly with single-ended timebases. Often there is a requirement for a symmetrical push-pull timebase. The simplest way to achieve this is to feed the output of a single-ended timebase into a phase splitter. In Fig. 11.6 (a) this is done by a conventional single-transistor phase-splitter. A more refined splitter is the long-tailed pair circuit shown in Fig. 11.6 (b). Finally a separate phase inverting transistor may be inserted. A typical sort of circuit with this last arrangement is given in Fig. 11.6 (c). Here Q1 is the 'switch' transistor for the Miller Integrator transistor Q2. The output from Q2 collector is fed via R_F to the base of the inverter transistor Q3, and an equal resistor R_F is used for feedback from the collector to the base of Q3. Thus Q3 acts as an operational phase inverter and the output at the Q3 collector is positive-going when the Q2 collector is negative-going. The symmetrical antiphase voltages at points A and B can be used to drive the horizontal plates of a cathode ray tube. The emitter resistor R_E common to Q2 and Q3 does not introduce degeneration into the Q2, Q3 circuit because the current through it remains constant, since the sum of the currents through Q2 and Q3 does not change. Sometimes a decoupling capacitor is used (as shown dotted) to eliminate switching transients.

FREE-RUNNING TIMEBASES

The timebases discussed so far have been mostly 'driven' by a train of input pulses. In practice timebases are often required to be free running. An illustration of such a circuit operating at about 10 kHz is shown in Fig. 11.7. In this the stage incorporating the n-p-n transistor Q1 is an astable blocking oscillator, which supplies the drive to the bootstrap integrator circuit Q2. The output from Q2 can be taken off directly at its emitter as a medium-impedance output or, if a higher load current is required, can be taken off at the output of the second buffer emitter follower Q3. From either output a voltage sweep is provided. Although the free running frequency of the blocking oscillator is about 10 kHz, its frequency can be varied over a range by means of the 2·5 kΩ variable resistor shown as 'frequency control'. Also it is possible to feed a train of synchronising input pulses into the base of Q1 to synchronise the timebase accurately to an external control frequency, provided the control frequency is slightly higher than the preset free-running

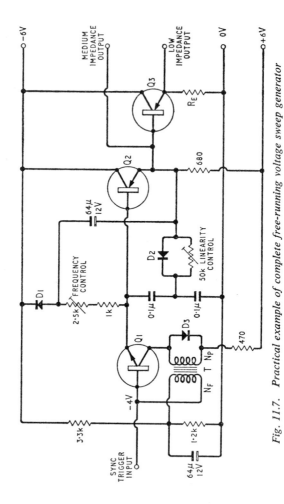

Fig. 11.7. Practical example of complete free-running voltage sweep generator

frequency of the blocking oscillator. The variable 50 kΩ resistor is provided to enable the linearity of the output sweep voltage to be optimised.

CURRENT TIMEBASES

Up to this point we have tacitly been considering voltage rather than current sweeps. In electromagnetic deflection systems, however, there is a specific requirement for a current rather than a voltage sweep. How a current sweep is achieved depends on whether the output transistor sees an inductive or a resistive load.

In television field timebases, which run at relatively slow speed (e.g. 50 Hz), the impedance of the field coils is for all practical purposes resistive. In this case, the linear current sweep is effected by using a transistor as a linear power amplifier as shown in Fig. 11.8 (a). A linear sweep (sawtooth) voltage is fed to the transistor base, and produces a sawtooth collector current in the deflection coil R_L. The inductance of R_L is so low compared with the load resistance at the slow sweep speed that the usual $\frac{L di}{dt}$ equation does not apply.

In line timebases, on the other hand (operating at greater than 10 kHz) the deflection coils are largely inductive at this high line frequency. To get a linear output current sweep it is necessary to feed a square wave to the transistor input as shown in Fig. 11.8 (b). In effect the transistor operates as a switch to transfer the supply

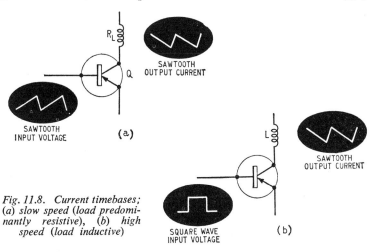

Fig. 11.8. Current timebases; (a) slow speed (load predominantly resistive), (b) high speed (load inductive)

voltage across the deflection coil for the sweep duration. The equation $V = L\frac{di}{dt}$ than gives a quasi-linear rising coil sweep current. For a fuller discussion of current timebases, readers should consult *Transistor Television Receivers* by T. D. Towers (Iliffe).

CONCLUSION

In this chapter we have had a look at the types of sweep generators that have been worked out with conventional transistors, but it should not be overlooked that over recent years new generations of semiconductor devices have grown up which in some cases have proved peculiarly suited to sweep generators. V.h.f. silicon epitaxial transistors enable timebases with very fast speeds to be devised, while field effect and metal-oxide-silicon type transistors with their exceptionally low leakage and high input resistance make possible timebases with very long periods. Other devices which have been widely used in timebase circuits are unijunctions, silicon controlled rectifiers and GTO's (gate turn-off units). In general, however, it will be found that these more refined circuits tend merely to be adaptations of the basic 'switch', 'Miller' or 'bootstrap' circuits described earlier.

APPENDIX A

Problems

Chapter 1

1. What is the highest current that can be switched by the 2N2369, the fast n-p-n switching transistor?
(*Answer:* 500 mA)

2. Explain, with reference to polarity, material, case outline, voltage ratings, current rating, power rating, frequency cut-off, current gain, switch-off time and collector capacitance, how a modern BFY50 can be used as a substitute for an obsolescent 2N1306.

3. (a) Arrange in order of physical size the four commonest metal can transistor case outlines: TO3, TO5/TO39, TO18 and TO66.
(b) Identify which of these has flexible leads and from a survey of the devices with these outlines in Appendix C work out the maximum current normally handled by flexible lead devices, i.e. with lead diameter less than 1 mm.
(*Answer:* (a) TO18–TO5/TO39–TO66–TO3; (b) TO18 = 800 mA, TO5/TO39 = 2 A)

4. Other things equal, what is the most usual polarity and semiconductor material used for general purpose switching transistors nowadays?
(*Answer:* n-p-n silicon)

5. How does the typical power dissipation of silicon switching diodes compare with germanium types?
(*Answer:* Germanium = 65–125 mW; Silicon = 250–1500 mW)

6. What is the essential difference between (a) 'astable', (b) 'bistable', and (c) 'monostable' multivibrators? Which multivibrator is self-oscillating?
(*Answer:* Astable)

180 Elements of Transistor Pulse Circuits

7. How does the 'paraphase' amplifier differ basically from the 'differential' amplifier?

8. If a phase inverter is followed by an emitter-follower and a further phase inverter, how does the phase of the final output compare with the phase of the initial input?
(*Answer:* In-phase)

9. Explain how the bootstrap capacitor in a high input impedance amplifier prevents the input bias resistor network from forming a low impedance shunt across the amplifier input. In Fig. 1.18, for a transistor with voltage amplification of 0·99 and an input bias resistor R3 = 10,000 ohms, what is the effective value of R3 due to the bootstrapping?
(*Answer:* 1 megohm)

10. Explain the basic action of the 'Miller' integrator. What is the 'effective' RC time constant of a Miller integrator with R = 100,000 ohms, C = 1 μF and an amplifier voltage gain A = ×100?
(*Answer:* 10 sec)

Chapter 2

1. Which is the commonest transistor configuration (common base, common emitter or common collector), and why?

2. (a) Outline the four basic h.f. compensation techniques used to extend the high frequency response of a pulse amplifier.
   (b) In the unbypassed emitter resistor arrangement of Fig. 2.5 (a), if R_E = 10 ohms, R_S = 490 ohms, h_{feo} = 100, and the bandwidth (with emitter decoupled) f_B = 200 kHz, what does the bandwidth become when the emitter bypass capacitor is removed?
(*Answer:* 600 kHz)

3. (a) How does the output (collector) capacitance of the transistor affect the high frequency response of an RC coupled amplifier?
   (b) What is the half-power frequency of a pulse amplifier stage with load resistance R_L = 500 ohms, a collector capacitance of 5 pF and a transistor current gain A_i = 50?
(*Answer:* 1·27 MHz)

4. (a) What factors mainly decide the low frequency response characteristics of a transistor pulse amplifier?
   (b) In Fig. 2.9 (b), what is the value of the emitter bypass capacitor required to give not more than 1 dB low frequency attenuation at 10 kHz in a pulse amplifier with R_S = 960 ohms and h_{feo} = 20?
(*Answer:* 1 μF)

Appendix A 181

5. What is the overall bandwidth for four cascaded identical stages in a pulse amplifier, where the bandwidth of a single stage is 1 MHz?
(*Answer:* 415 kHz)

Chapter 3

1. Explain the operating principle of a free-running astable multivibrator.

2. In a synchronised astable multivibrator, is the free-running frequency higher or lower than the forced frequency? Explain.

3. What main factors prevent you indefinitely reducing the pulse repetition frequency of an astable multivibrator?

4. What is the approximate pulse repetition frequency of a symmetrical collector-coupled astable multivibrator with cross-coupling capacitors $C = 1\ \mu F$ and base feed resistors $R = 100{,}000$ ohms?
(*Answer:* 7 Hz)

5. Frequency division of more than 10 times is not commonly obtained using an astable multivibrator. Explain.

Chapter 4

1. Explain the principle of operation of a monostable multivibrator.

2. What is the approximate pulse length in the monostable of Fig. 4.1 if $C = 1\ \mu F$ and $R_{B2} = 150{,}000$ ohms?
(*Answer:* 0·1 sec)

3. Explain the basic differences between collector- and emitter-coupled transistor monostable multivibrators.

4. Explain the operation of a 'normally on' complementary symmetry monostable such as Fig. 4.7.

5. Outline the main differences between the 'normally off' complementary symmetry monostable of Fig. 4.8 and the 'normally on' type of Fig. 4.7.

Chapter 5

1. Explain with reference to Fig. 5.1 the mechanism of switch over between states of a bistable multivibrator.

182 Elements of Transistor Pulse Circuits

2. For bistable triggering purposes, how do you obtain a short pulse coincident with the leading edge of a rectangular pulse?

3. Estimate the maximum practical pulse repetition rate achievable with a bistable multivibrator using BC107 transistors, having $f_{hfb} = 100$ MHz and $h_{fe} = 225$.
(*Answer:* 4·5 MHz)

4. Outline the essential differences between (a) base, (b) collector and (c) hybrid triggering of a bistable multivibrator.

5. (a) Explain the purpose of the 'commutating' capacitor in the multivibrator.
(b) What is the approximate value of commutating capacitor required for a transistor with $h_{fe} = 50$, $f_{hfb} = 100$ MHz and base input drive resistor $R_B = 10,000$ ohms?
(*Answer:* 8 pF)

Chapter 6

1. Outline the essential differences between high-pass (differentiator) and low pass (integrator) RC filter circuits.

2. What is the 'lower 3 dB frequency' of a high-pass RC filter with $R = 10,000$ ohms and $C = 0.01$ μF?
(*Answer:* 1,600 Hz)

3. Calculate the percentage droop on a steady state square wave train of pulse repetition frequency $f = 1,000$ Hz passed through a high-pass filter with $R = 10,000$ ohms and $C = 1$ μF.
(*Answer:* 5%)

4. What is the upper 3 dB frequency of a low pass RC filter with $R = 10,000$ ohms and $C = 0.01$ μF?
(*Answer:* 1,600 Hz)

5. In an overdriven transistor amplifier with a rail voltage of 10 V and a collector load resistor of 1,000 ohms and with no emitter resistor, between what levels does the output level switch?
(*Answer:* 10 V—approx. 0 V)

Chapter 7

1. (a) What are the limitations to the use of a simple two-diode pump for generating a linear-output staircase waveform?

Appendix A   183

(b) In Fig. 7.1 (a), suggest suitable values for capacitors C1 and C2 when the pump is driven from a pulse generator with source resistance 1,000 ohms providing 100 μs pulses.
(*Answer:* C1 = 250 nF, C2 = 0·005 μF)

2. Outline the principle of operation of the transistor pump.

3. (a) What advantage has the Miller integrator diode pump over the simple two-diode pump?
(b) When an input pulse of 6 V is fed into a Miller diode pump through a series capacitor C1 = 0·001 μF and with a feedback capacitor C2 = 0·03 μF, what is the output voltage increment?
(*Answer:* 0·2 V)

4. (a) Describe the basic operation of the transistor Schmitt trigger of Fig. 7.4.
(b) Between what levels does the output switch if R1 = 1,000 ohms, R2 = 100 ohms and V_{cc} = 11 V?
(*Answer:* 11 V and 1 V)

5. Describe two common applications of the Schmitt trigger.

Chapter 8

1. What are the three basic feedback arrangements for the transistor blocking oscillator?

2. Describe the full cycle of a blocking oscillator firing and recovery, using the example of Fig. 8.1 (a).

3. (a) Discuss the various control mechanisms that may be used in blocking oscillator design to define the pulse length.
(b) What is the pulse length for the circuit of Fig. 8.3 (a), when L_C = 1 mH, N_B/N_C = 1/5, $(R_B + R_E)$ = 500 ohms and R_O = 5,000 ohms?
(*Answer:* 0·2 μs)

4. Show how the non-saturated blocking oscillator differs from the saturated version.

5. (a) What restricts the recovery time of a blocking oscillator?
(b) In a blocking oscillator where the transformer primary has an inductance of 1 mH and a resistance of 1,000 ohms reflected into the primary from the load and circuit resistances, what is the approximate recovery time?
(*Answer:* 4 μs)

184 *Elements of Transistor Pulse Circuits*

Chapter 9

1. (a) Distinguish between a 'transmission' and a 'logical' gate.
  (b) What is the 'gain' A $(=V_O/V_G)$ of the transmission gate of Fig. 9.1 (c) when $R_G = R_S = 10,000$ ohms and $R_O = 1,000$ ohms? (*Answer:* $A = 1/7$)

2. Outline the advantages of transistor transmission gates as compared with diode gates.

3. Explain the operation of basic OR and AND diode logic gates.

4. Give illustrative basic examples of transistor NOT, OR, AND and NOR gates.

5. Explain the functional operation of an 'anti-coincidence' logical gate.

Chapter 10

1. Explain the difference between the frequency-measuring and time-measuring modes of operation of a counter/timer.

2. What is a 'DCU'? Explain how found bistable multivibrators ('binaries') can be connected with feedback to form a DCU.

3. Describe a typical master control oscillator for a counter/timer.

4. What are the functions of a control unit in a counter/timer?

Chapter 11

1. (a) Explain (i) the 'displacement error', (ii) 'slope error' and (iii) 'transmission error' in a nominally linear sweep output, and state the approximate relationship between them.
  (b) In charging a 1 μF capacitor up to 10 V through a 10,000 ohm resistor from a 100 V d.c. supply, what is the displacement error? (*Answer:* 1·25%)

2. (a) Describe the basic circuit operation of a transistor Miller sweep generator.
  (b) How much is the linearity of a Miller sweep with a voltage gain of 100 better than a simple RC charging sweep?
(*Answer:* 100 times)

3. (a) Explain how the bootstrap sweep generator operates, particularly in relation to its difference from the Miller sweep.
  (b) How much better is the linearity of a bootstrap sweep with an

emitter-follower gain of 0·98 than the corresponding simple RC sweep? (*Answer:* 50 times)

4. Illustrate the use of a blocking oscillator as a regenerative sweep generator.

5. How can an astable multivibrator be connected up to provide a periodic sweep output?

APPENDIX B

Bibliography

(*for more detailed treatments of pulse circuit types*)

AMOS, S. W., *Principle of transistor circuits*, Iliffe-Butterworth (1969).
BLITZER, R., *Basic pulse circuits*, McGraw-Hill (1964).
BUDINSKY, J., *Techniques of transistor switching circuits*, Iliffe-Butterworth (1968).
COMER, D. T., *Large signal transistor circuits*, Prentice-Hall (1967).
DELHOM, L. A., *Design and applications of transistor switching circuits*, McGraw-Hill (1968).
GHAUSI, M. S., *Electronic circuits*, Van Nostrand (1971).
GILLIE, A. C., *Pulse and logic circuits*, McGraw-Hill (1968).
HAWKINS, J. K., *Circuit design of digital computers*, Wiley (1968).
HURLEY, R. B., *Transistor logic circuits*, Wiley (1961).
LITTAUER, R., *Pulse electronics*, McGraw-Hill (1965).
LITTWIN, S., *Pulse generators in industrial electronics*, Iliffe-Butterworth (1965).
MILLMAN, J. AND TAUB, H., *Pulse, digital and switching circuits*, McGraw-Hill (1965).
MITCHELL, B. B., *Semiconductor pulse circuits with experiments*, Holt, Rinehart and Winston (1970).
STRAUSS, L., *Wave generation and shaping*, McGraw-Hill (1970).
PETTIT, J. and MCWHORTER, M., *Electronic switching, timing and pulse circuits*, McGraw-Hill (1970).
THOMAS, H. E., *Handbook of transistors, semi-conductors, instruments and micro-electronics*, Prentice-Hall (1969).

Transistor Specifications and Substitutions

(Detailed data for devices referred to in text)

Type	*Pol & Mat	*Case	V_{ceo} max	V_{ebo} max	I_c max	P_{tot} max	F_t min	$H_{fe} @ I_c$ min	T_{off} max	C_{ob} max	Substitute
2N406	PG	TO1	20 V	3 V	35 mA	0·15 W	300 K	35 @ 1 mA	10 μs	70 pF	BC177
2N647	NG	TO1	25 V	12 V	0·1 A	0·1 W	1 M	50 @ 50 mA	5 μs	30 pF	BC107
2N703	NS	TO18	25 V	5 V	50 mA	0·3 W	70 M	40 @ 10 mA	100 ns	6 pF	2N2369
2N706	NS	TO18	20 V	3 V	0·1 A	0·3 W	200 M	20 @ 10 mA	75 ns	6 pF	2N2369
2N709	NS	TO18	6 V	4 V	0·2 A	0·3 W	600 M	20 @ 10 mA	15 ns	3 pF	2N2475
2N834	NS	TO18	30 V	5 V	0·2 A	0·3 W	350 M	25 @ 10 mA	75 ns	4 pF	2N2369
2N967	PG	TO18	12 V	2 V	0·1 A	0·15 W	300 M	40 @ 10 mA	120 ns	5 pF	2N2894
2N1305	PG	TO5	20 V	25 V	0·3 A	0·15 W	5 M	40 @ 10 mA	2 μs	20 pF	BC177
2N1306	NG	TO5	20 V	25 V	0·3 A	0·15 W	5 M	40 @ 10 mA	2 μs	20 pF	BC107
2N1533	PG	TO3	60 V	60 V	10 A	30 W	100 K	15 @ 6 A	50 μs	300 pF	BDX18
2N1990	NS	TO39	100 V	3 V	1 A	0·6 W	100 M	20 @ 30 mA	5 μs	10 pF	BSW66
2N2222	NS	TO18	30 V	5 V	0·8 A	0·5 W	250 M	30 @ 0·5 A	300 ns	8 pF	—
2N2297	NS	TO5	35 V	7 V	1 A	0·8 W	60 M	15 @ 1 A	1 μs	12 pF	BFY50
2N2369	NS	TO18	15 V	4 V	0·5 A	0·36 W	500 M	20 @ 0·1 A	18 ns	4 pF	BSX20
2N2894	PS	TO18	12 V	4 V	0·2 A	0·36 W	400 M	25 @ 0·1 A	90 ns	6 pF	—
2N2907	NS	TO18	40 V	5 V	0·6 A	0·4 W	200 M	30 @ 0·5 A	100 ns	8 pF	—
2N2924	NS	TO98	25 V	5 V	0·1 A	0·2 W	—	150 @ 2 mA	—	12 pF	—
2N3054	NS	TO66	55 V	7 V	4 A	25 W	800 K	25 @ 0·5 A	5 μs	—	—
2N3055	NS	TO3	60 V	7 V	15 A	115 W	800 K	20 @ 4 A	15 μs	250 pF	—
2N3426	NS	TO5	12 V	4 V	1 A	0·6 W	200 M	20 @ 10 mA	35 ns	25 pF	—
2N3442	NS	TO3	140 V	7 V	15 A	117 W	800 K	20 @ 3 A	10 μs	—	—

* N = n-p-n, P = p-n-p, G = germanium, S = silicon; drawings for case outlines are given at the end of this Appendix.

188 Elements of Transistor Pulse Circuits

Type	*Pol & Mat	*Case	V_{ceo} max	V_{ebo} max	I_c max	P_{tot} max	F_t min	H_{fe} @ I_c min	T_{off} max	C_{ob} max	Substitute
2N3497	PS	TO18	120 V	4 V	0·1 A	0·4 W	150 M	40 @ 10 mA	1 μs	6 pF	—
2N3725	NS	TO5	50 V	6 V	0·5 A	0·8 W	300 M	60 @ 0·1 A	80 ns	10 pF	—
2N3763	PS	TO5	60 V	5 V	1·5 A	1 W	150 M	20 @ 1 A	115 ns	15 pF	—
2N3792	PS	TO3	80 V	7 V	10 A	150 W	4 M	30 @ 3 A	1 μs	—	—
2N4036	PS	TO5	65 V	7 V	1 A	1 W	60 M	20 @ 0·1 A	700 ns	30 pF	—
2N4207	PS	TO18	6 V	4 V	50 mA	0·3 W	650 M	50 @ 1 mA	25 ns	3 pF	—
2N4390	NS	TO18	120 V	6 V	0·2 A	0·38 W	50 M	20 @ 20 mA	1·3 μs	6 pF	BFR25
2N5886	NS	TO3	80 V	5 V	20 A	200 W	4 M	20 @ 10 A	1·8 μs	500 pF	—
AC125	PG	TO1	12 V	10 V	0·1 A	0·3 W	1·3 M	50 @ 2 mA	2 μs	50 pF	BC177
AC127	NG	TO1	12 V	10 V	0·5 A	0·34 W	1·5 M	50 @ 20 mA	2 μs	100 pF	2N2222
AC128	PG	TO1	16 V	10 V	1 A	0·27 W	1 M	60 @ 0·3 A	2 μs	150 pF	2N4036
AC176	NG	TO1	32 V	5 V	1 A	0·7 W	1·5 M	50 @ 0·5 A	2 μs	180 pF	BFY50
AD161	NG	—	20 V	10 V	3 A	4 W	800 K	80 @ 0·5 A	5 μs	250 pF	2N3054
AD162	PG	—	20 V	10 V	3 A	6 W	6 M	80 @ 0·5 A	5 μs	175 pF	BDX14
ASY27	PG	TO5	15 V	20 V	0·3 A	0·15 W	6 M	50 @ 20 mA	2·2 μs	16 pF	2N2894
ASY29	NG	TO5	20 V	20 V	0·3 A	0·15 W	125 K	50 @ 20 mA	1·3 μs	16 pF	2N2369
ASZ16	NG	TO3	60 V	20 V	10 A	30 W	300 M	35 @ 6 A	50 μs	300 pF	BDX18
ASZ21	PG	TO18	15 V	2 V	50 mA	0·12 W	150 M	30 @ 10 mA	115 ns	5 pF	2N2894
BC107	NS	TO18	45 V	6 V	0·2 A	0·3 W	40 M	110 @ 2 mA	1 μs	5 pF	—
BC125	NS	TO105	30 V	5 V	0·6 A	0·3 W	80 M	30 @ 0·15 A	200 ns	25 pF	—
BC126	PS	TO105	30 V	5 V	0·6 A	0·3 W	30 M	80 @ 0·15 A	150 ns	12 pF	—
BC154	PS	TO106	40 V	5 V	0·1 A	0·2 W	50 M	160 @ 10 mA	200 ns	10 pF	—
BC160	PS	TO39	40 V	5 V	1 A	0·65 W	100 M	40 @ 0·1 A	2 μs	30 pF	—
BC177	PS	TO18	45 V	5 V	0·2 A	0·3 W	150 M	30 @ 10 μA	150 ns	8 pF	—
BC182L	NS	TO92	50 V	5 V	0·2 A	0·3 W	200 M	100 @ 2 mA	100 ns	5 pF	—
BC213L	PS	TO92	30 V	5 V	0·2 A	0·3 W	130 M	60 @ 2 mA	150 ns	10 pF	—
BC257	PS	TO92	45 V	5 V	0·1 A	0·3 W		70 @ 2 mA		6 pF	—

Appendix C 189

Type	*Pol & Mat	*Case	V_{ceo} max	V_{ebo} max	I_c max	P_{tot} max	F_t min	H_{fe} @ I_c min	T_{off} max	C_{ob} max	Substitute
BD135	NS	TO126	45 V	5 V	0·5 A	6·5 W	50 M	40 @ 0·15 A	—	—	—
BD136	PS	TO126	45 V	5 V	0·5 A	6·5 W	50 M	40 @ 0·15 A	—	—	—
BD213	NS	TOP3	45 V	7 V	15 A	90 W	3 M	25 @ 5 A	—	—	—
BD214	PS	TOP3	45 V	7 V	15 A	90 W	3 M	25 @ 5 A	—	—	—
BD595	NS	TOP66	45 V	5 V	8 A	55 W	3 M	25 @ 3 A	—	—	—
BD596	PS	TOP66	45 V	5 V	8 A	55 W	3 M	25 @ 3 A	—	—	—
BDX11	NS	TO3	140 V	7 V	15 A	117 W	800 K	20 @ 3 A	10 μs	—	—
BDX14	PS	TO66	60 V	7 V	4 A	29 W	800 K	25 @ 0·5 A	10 μs	—	—
BDX18	PS	TO3	60 V	7 V	15 A	117 W	4 M	20 @ 4 A	1 μs	—	—
BDY20	NS	TO3	60 V	7 V	15 A	115 W	1 M	20 @ 4 A	5 μs	—	—
BDY57	NS	TO3	80 V	10 V	25 A	175 W	10 M	20 @ 4 A	1 μs	—	—
BDY90	NS	TO3	100 V	6 V	15 A	40 W	35 M	30 @ 5 A	1 μs	—	—
BFR39	NS	TO92	80 V	5 V	2 A	0·8 W	100 M	50 @ 0·1 A	500 ns	10 pF	—
BFR79	PS	TO92	80 V	5 V	2 A	0·8 W	100 M	50 @ 0·1 A	500 ns	10 pF	—
BFS95	PS	TO39	35 V	6 V	1 A	1 W	40 M	70 @ 0·15 A	1·5 μs	20 pF	—
BFY50	NS	TO5	35 V	6 V	1 A	0·8 W	60 M	15 @ 1 A	0·5 μs	12 pF	—
BLX41	PS	TO39	100 V	10 V	2 A	1 W	—	40 @ 0·5 A	5 μs	—	—
BSV68	PS	TO18	100 V	6 V	0·1 A	0·36 W	50 M	30 @ 25 mA	0·5 μs	5 pF	—
BSW19	NS	TO18	30 V	5 V	0·1 A	0·22 W	300 M	40 @ 10 mA	800 ns	6 pF	—
BSW25	PS	TO18	12 V	4 V	0·2 A	0·36 W	800 M	40 @ 30 mA	75 ns	5 pF	—
BSW64	PS	TO18	40 V	6 V	0·8 A	0·5 W	300 M	100 @ 0·15 A	200 ns	8 pF	—
BSW66	NS	TO39	100 V	6 V	1 A	0·8 W	40 M	30 @ 0·5 A	2 μs	35 pF	—
BSX12	NS	TO5	12 V	4 V	1 A	0·6 W	450 M	20 @ 10 mA	25 ns	15 pF	—
BSX20	NS	TO18	15 V	4 V	0·5 A	0·36 W	500 M	10 @ 0·1 A	18 ns	4 pF	2N2369
BSX21	NS	TO18	80 V	5 V	0·2 A	0·3 W	60 M	20 @ 4 mA	200 ns	5 pF	—
BSX 36	PS	TO18	40 V	5 V	0·5 A	0·36 W	100 M	40 @ 10 mA	100 ns	8 pF	—

* N = n-p-n, P = p-n-p, G = germanium, S = silicon; drawings for case outlines are given at the end of this Appendix.

Type	*Pol & Mat	*Case	V_{ceo} max	V_{ebo} max	I_c max	P_{tot} max	F_t min	H_{fe} @ I_c min	T_{off} max	C_{ob} max	Substitute
BSX59	NS	TO5	45 V	5 V	1 A	0.8 W	250 M	25 @ 0.5 A	60 ns	10 pF	—
BSY39	NS	TO18	15 V	5 V	0.2 A	0.3 W	350 M	40 @ 10 mA	25 ns	5 pF	—
BSY95A	NS	TO18	15 V	5 V	0.2 A	0.3 W	200 M	50 @ 10 mA	100 ns	6 pF	2N2369
OC42	PG	TO1	15 V	6 V	50 mA	50 mW	6 M	30 @ 50 mA	1 μs	13 pF	BC177
OC71	PG	TO1	18 V	6 V	10 mA	125 mW	200 K	20 @ 3 mA	5 μs	20 pF	BC177
OC72	PG	TO1	16 V	10 V	0.1 A	165 mW	200 K	30 @ 10 mA	5 μs	35 pF	BC177

* N = n-p-n, P = p-n-p, G = germanium, S = silicon; drawings for case outlines are given at the end of this Appendix.

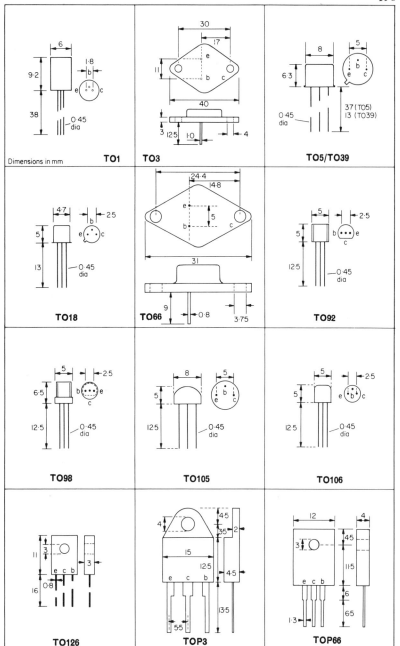

APPENDIX D

Diode Specifications and Substitutions

(Detailed data for devices referred to in text)

Type	*Mat and Case	P_{tot} max	I_f max	V_r max	V_f @ I_f max	I_r @ V_r max	T_{rr} max	C_d max	Substitute
1N914	S DO7	250 mW	225 mA	100 V	1 V @ 10 mA	25 nA @ 20 V	4 ns	4 pF	1N4148
1N4001	S DO41	1·5 W	3 A	50 V	1·1 V @ 1 A	10 µA @ 50 V	10 µs	30 pF	1N4002
1N4002	S DO41	1·5 W	3 A	100 V	1·1 V @ 1 A	10 µA @ 100 V	10 µs	30 pF	1N4003
1N4003	S DO41	1·5 W	3 A	200 V	1·1 V @ 1 A	10 µA @ 200 V	10 µs	30 pF	1N4004
1N4004	S DO41	1·5 W	3 A	400 V	1·1 V @ 1 A	10 µA @ 400 V	10 µs	30 pF	1N4005
1N4005	S DO41	1·5 W	3 A	600 V	1·1 V @ 1 A	10 µA @ 600 V	10 µs	30 pF	1N4006
1N4006	S DO41	1·5 W	3 A	800 V	1·1 V @ 1 A	10 µA @ 800 V	10 µs	30 pF	1N4007
1N4007	S DO41	1·5 W	3 A	1 kV	1·1 V @ 1 A	10 µA @ 1 kV	10 µs	30 pF	—
1N4009	S DO7	250 mW	100 mA	25 V	1 V @ 30 mA	100 nA @ 25 V	2 ns	4 pF	1N4154
1N4148	S DO35	500 mW	225 mA	100 V	1 V @ 10 mA	25 nA @ 25 V	4 ns	4 pF	1N914
1N4151	S DO35	500 mW	450 mA	75 V	1 V @ 50 mA	50 nA @ 50 V	2 ns	2 pF	1N3604
1N4376	S DO7	250 mW	100 mA	20 V	1·1 V @ 50 mA	100 nA @ 10 V	0·8 ns	1 pF	1N4244
AAY11	G DO7	125 mW	150 mA	90 V	1 V @ 5 mA	65 µA @ 50 V	1 µs	—	—
AAY21	G DO7	66 mW	50 mA	15 V	0·8 V @ 10 mA	10 µA @ 5 V	12 ns	1·2 pF	—
AAY30	G DO7	84 mW	400 mA	50 V	1 V @ 150 mA	15 µA @ 10 V	—	1 pF	—
AAZ13	G DO7	100 mW	100 mA	8 V	1 V @ 30 mA	25 µA @ 8 V	—	2 pF	—
AAZ15	G DO7	75 mW	250 mA	100 V	1·1 V @ 250 mA	4 µA @ 10 V	700 ns	2 pF	—
AAZ17	G DO7	75 mW	250 mA	75 V	1·1 V @ 250 mA	60 µA @ 10 V	350 ns	2 pF	—
AAZ18	G DO7	66 mW	300 mA	20 V	0·8 V @ 300 mA	15 µA @ 10 V	70 ns	1·5 pF	—

Type	*Mat and Case	P_{tot} max	I_f max	V_r max	$V_f @ I_f$ max	$I_r @ V_r$ max	T_{rr} max	C_d max	Substitute
BAX12	S SPEC	600 mW	800 mA	120 V	1 V @ 200 mA	—	50 ns	35 pF	—
BAX13	S SPEC	300 mW	150 mA	50 V	1 V @ 20 mA	25 nA @ 10 V	4 ns	3 pF	1N4148
BAX16	S SPEC	440 mW	300 mA	150 V	1.3 V @ 100 mA	25 nA @ 50 V	120 ns	10 pF	—
BAV10	S DO35	390 mW	600 mA	60 V	1.3 V @ 500 mA	100 nA @ 60 V	6 ns	2.5 pF	AAZ12
OA10	G	—	1 A	30 V	0.6 V @ 100 mA	600 µA @ 30 V	2 µs	20 pF	AAZ15
OA47	G DO7	110 mW	150 mA	25 V	1.1 V @ 150 mA	15 µA @ 10 V	70 ns	3.5 pF	OA95
OA91	G DO7	125 mW	150 mA	115 V	3.3 V @ 30 mA	11 µA @ 10 V	—	—	OA91
OA95	G DO7	125 mW	150 mA	115 V	2.6 V @ 30 mA	7 µA @ 10 V	—	—	1N4148
OA200	S DO7	250 mW	250 mA	50 V	1.2 V @ 30 mA	100 nA @ 50 V	3 µs	25 pF	—
OA202	S DO7	250 mW	250 mA	150 V	1.2 V @ 30 mA	100 nA @ 150 V	3 µs	25 pF	—

* = silicon, G = germanium; drawings for case outlines are given below at the end of this Appendix.

Index

Adder, brief review, 19
Amplifier
  bootstrap, 24
  buffer, 160
  differential, 14
  high-input-impedance, 23–25
  linear pulse. *See* Linear pulse amplifiers
  operational, 15
  over-driven, 97
  paraphase, 13
  signal gate buffer, 160

Balanced inverter, 12
Blocking oscillators, 113–133
  applications 129
  as frequency divider, 131
  as timebase, 171
  astable, 103, 175
    beta-limited and non-saturated, 131
    operation of, 115
    time constant for pulse repetition, 126
  basic circuit, 118
  beta-limitation, 116, 119–122
  brief review of, 25
  circuit arrangements, 113
  core saturation, 116, 122, 123
  d.c. biased transistor, 126
  delay-line-limitation, 116
  design requirements, 118
  feedback arrangements, 113
  free-running, 126

Blocking oscillators *continued*
  monostable, 119
    saturated collector-base feedback, 119–122, 131
    saturated collector-emitter feedback, 122
    saturated emitter-base feedback, 122
  non-saturated, 123
  operating principle, 115
  output take-off, 129
  practical examples of circuits, 131
  pulse drop, 118
  RC-limitation, 116
  recovery time, 125
  reverse spike of, 125
  saturated or non-saturated transistor operation, 118
  s.c.r.-firing monostable, 131
  switch-off-limitation, 116
  switching cycle detail, 117
  timing mechanisms, 116, 118
  transformer feedback mode, 119
  trigger-pulse controlled, 126
  tuned-circuit limitation 116
Bootstrap amplifier, 24
Bootstrap circuit, 21
Bootstrap timebase, 168, 173
Buffer amplifiers, 160

Capacitance meter, 105
Capacitor in Schmitt trigger, 108
Capacitor-compensation for Eccles-Jordan multivibrator, 80
Chopper, transistor, 140

196 *Index*

Clamper, unbiased, 96–97
Clipper circuits, 95
  capacitor-coupled, 96
Collector capacitance effect, 38
Compound emitter follower, 23
Counter/timers, 148–163
  basic principle, 148
  binary element, 151
  block diagram, 148–150
  buffer amplifier and inverter, 160
  components of, 150
  counter in, 151
  decade counter units, 150, 152
  gate control units, 150–151, 158
  input pulse shaper, 150, 154
  latch buffer inverter, 160
  master control oscillator, 150, 157
  power supply, 150
  readout display, 150, 152
  signal gate, 150, 154
    buffer amplifier, 160
  start reset delay unit, 159–162
  timebase, 150, 157
  timebase dividers, 150, 158
Coupling, interstage, 31

Darlington pair, 56, 162
D.c. level detector, 112
Delay time, 117
Differential amplifier, 14
Differentiator, 19, 85, 88
Diode AND gate, 142
Diode clippers, 95
Diode OR gate, 141
Diode pump, 99, 100–103
  applications, 104–105
Diodes,
  as non-linear passive waveshaping elements, 93
  case outlines, 193
  germanium, 9
  germanium, gold-bonded, 9
  germanium, junction, 10
  germanium, point contact, 9
  in transmission gates, 135–138
  silicon, 10
  standard devices, 192
  substitution of, 10
  switching, 8
Displacement error, 172,173
Double clipper, 96
Double-diode clipper 98

Eccles-Jordan bistable multivibrators. *See* Multivibrators
Emitter-coupled binary, 105
Emitter follower, brief review, 12

Equivalent circuits of linear pulse amplifiers, 36
Exponential waveforms, 84

Filters
  high-pass RC, 85–89
  low-pass RC, 89–93
  low-time-constant high-pass RC, 89
Flip, 52
Flyback, 164,165
Fourier Synthesis, 84
Frequency characteristics of linear pulse amplifiers, 33
Frequency control, 175
Frequency divider
  blocking oscillator as, 131
  synchronised astable multivibrator, 56
Frequency meter, 104, 148–153
Frequency-voltage converter, 105

Gain-bandwidth product, 36, 37, 42
Gates, 134–147
  AND, 141, 142, 144, 145
  AND-NOT, 146
  ANTI-COINCIDENCE, 146
  building block, 145
  COINCIDENCE, 142
  DOWN, 141, 142, 143
  EXCLUSIVELY-OR, 146
  four-diode, 136
  in counter/timers, 150
  INHIBITOR, 146
  linear, 134
  logical, 134
    practical aspects, 146
    supplementary transfer, 146
    symbols for, 145
  NAND, 146
  NOR, 145
  NOT, 142, 144, 145
  NOT-AND, 146
  OR, 141, 142, 144–146
  single-diode, 135
  six-diode, 137
  transmission, 134
    using diodes, 135–138
    using transistors, 138
  two-diode, 135
  UP, 141, 142, 145, 146

High-frequency compensation, 38–43
Hybrid-pi circuit, 36

Integrator, 17, 89, 92
Interstage coupling, 31
Inverter, 142
  latch buffer, 160
Inverter transistor, 175

Index 197

Leakage current, transistor, 76
Limiters, 95.
Linear pulse amplifiers, 30–47
  collector capacitance effect, 38
  current gain without frequency compensation, 37
  equivalent circuits, 36
  frequency characteristics, 33
  gain-bandwidth product, 36, 37, 42
  general design problems, 30
  high-frequency compensation, 38–43
  high-level, 46
  interstage coupling, 31
  low-frequency compensation, 44
  multistage, 45
  negative feedback high-frequency compensation, 38
  positive feedback high-frequency compensation, 42
  pulse characteristics, 32
  RC-coupled, 31, 36
  series-inductance high-frequency compensation, 42
  shunt-inductance high frequency compensation, 40
  testing bandwidth characteristic, 43
  transistors in, 31, 45
  typical uncompensated stage, 38
Linear ramp, 91–92
Linear sweep generators, review of, 20–23
Low-frequency compensation, 44

Miller-integrator, 22, 175
Miller-integrator pump, 103
Miller-integrator timebase 167, 173,
Multivibrator,
  astable 26, 48–58
    description of, 50
    high-frequency limitations, 54
    low-frequency limitations, 55
    operating principle, 50
    oscillation frequency, 52
    output, 54
    practical design, 53
    synchronised frequency divider, 56
  bistable, 26, 48, 49
    uses of, 71
  brief review of 26
  definition of, 48–49
  Eccles-Jordan bistable, 71–83
    base cut-off bias resistor, 75
    base supply voltage, 75
    basic circuit, 72
    basic d.c. design, 73–77
    collector-base cross-coupling resistor circuit, 74

Eccles-Jordan bistable *continued*
    collector current in 'on' transistor, 74
    d.c. design of single-power-supply circuit, 76
    ouput voltage swing, 73–74
    practical examples of, 82
    steering circuits, 78
    switching between stable states, 77
    switching speed, 80
    trigger circuits, 78–83
  free-running, 56
  history of, 49
  monostable, 28, 48, 49, 59–70, 162
    basic collector-coupled, 60
    collector coupled, 60–63
    complementary-symmetry, 68
    description of, 59–60
    emitter-coupled, 60, 67
    typical high-speed, 64
    typical low-speed collector-coupled, 65
    typical medium-speed collector-coupled, 63
  regenerative sweep generators, 171
  symmetrical, 53

Negative feedback high-frequency compensation, 38
NORBIT system of switching blocks, 145

On/off switch, 134, 135
Operational amplifier, 15

Paraphase amplifier, 13
Phantastron, 166
Phase inverter, 11
Phase-inverting transistor, 175
Phase shifter, 17
Phase splitter 12, 175
Positive feedback high-frequency compensation, 42
Pulse characteristics of linear pulse amplifiers, 32
Pulse reshaper or restorer, 112
Pulse reshaping, 131
Pulses, 84
Pumps, 100–105

Ramp generator. *See* Timebases
Ramp input voltage, 88
Ramp waveforms, 84
RC-coupling, 31, 36
Readout display, 150–152
Rectangular pulse, 84, 85, 87, 90
Regenerative switching circuits, 25–29
Relax, 52

198 Index

Reverse spike of blocking oscillator, 125
Rise time, 90
RL linear passive networks, 93

Sawtooth generator. *See* Timebases
Sawtooth wave, 84, 165
Scale changer, 16
Schmitt trigger, 100, 105–112
  applications, 111
  brief review of, 28
  circuit arrangement, 105
  circuit refinements, 108
  design outline method, 106–108
  practical examples, 109
Scoop counter, 104
Series-inductance high-frequency compensation, 42
Shunt capacitor, 89
Shunt-inductance high-frequency compensation, 40
Sign changer, 16
Sine wave 84, 89–90, 98
Slicer, 96
Slope error, 172
Square wave generator, 53
Square waves, 84, 87, 165
Squarer, 111
Staircase ramp waveforms, 99
Staircase waveform, 100–103, 104
Steering circuits for Eccles-Jordan multivibrator, 78
Step function, 84, 87, 88,
Step voltage, 85, 90
Storage counter, 104
Sweep, 164
Sweep generator. *See* Timebases
Switch, on/off, 134–135
Switching blocks NORBIT system of 145
Switching circuits, regenerative, 25–29
Switching design for Eccles-Jordan multivibrator, 77
Switching speed of transistor, 55

Timebases, 164–178
  basic trigger, 170
  blocking oscillator, as, 171
  bootstrap, 168, 173
  current, 177
  driven non-regenerative, 165
  free-running, 175
  linear, 164
    output current, 177
  linearity considerations, 172
  Miller-integrator, 167, 173
  multivibrator regenerative, 171
  new devices in 178

Timebases *continued*
  non-regenerative linear, 165
  output, 164
  single-ended, 175
  single sided, 164
  symmetrical, 175
    push-pull, 164, 175
  television field, 177
  transistor-switch, 165
  uses of, 164
  voltage, 177
Timers. *See* Counter/timers
Transistor,
  as triode, 138
  case outlines, 7, 191
  choosing right type, 2
  germanium, npn types, 3
  germanium, pnp types, 3
  high voltage types, 5
  in linear pulse amplifiers, 31, 45
  in transmission gates, 138
  inverter, 175
    phase-inverting, 175
  silicon v. germanium types, 6
  silicon, npn types, 4
  silicon, pnp types, 5
  specifications, 8
  standard devices, 187
  substitutions, 8
  switching speed, 55
  voltage ratings, 53
Transistor AND gate, 144
Transistor chopper, 140
Transistor leakage currents, 76
Transistor OR gate, 142
Transistor pump, 99, 103
Transistor-switch timebase, 165
Transmission error, 88
Trigger circuits for Eccles-Jordan multivibrator, 78–83
Trigger-pulse controlled blocking oscillator, 126–128
Trigger sweep generators, 170
Triode, transistor as, 138

Voltage ratings of transistors, 53

Waveform shaping, 84–99
  linear, 84–93
  non-linear active, 97
  non-linear passive, 93–97
  practical use of, 98
Waveforms
  examples of common, 84
  exponential, 84
  ramp, 84
  sawtooth, 84, 165

Zero cross-over detector, 112